GL-121 & 122

STRATIGRAPHY, PALAEONTOLOGY AND PETROLOGY

Geology : Paper-I [2 Credits] & II [2 Credits]
For
First Year B.Sc. Semester-II
As Per New Syllabus of CBCS Pattern
From June - 2019

Dr. U. D. KULKARNI
Principal,
M. J. College
Jalgoan.

Dr. I. A. KHAN
Associate Professor, (Retd.)
Dept. of Geology and Petroleum Technology,
Nowrosjee Wadia College,
Pune.

Dr. M. V. BARIDE
Associate Professor, (Retd.)
Department of Geology,
Z.B. Patil College,
Dhulia.

Dr. A. B. CHAKRANARAYAN
Vice-Principal (Retd.)
Fergusson College,
Pune.

N5000

F.Y.B.Sc. GEOLOGY　　　　　　　　　　　　　　　　**ISBN 978-93-89825-41-1**

First Edition　:　January 2020
©　　　　　:　Authors

Published By:
NIRALI PRAKASHAN
Abhyudaya Pragati, 1312, Shivaji Nagar
Off J.M. Road, PUNE – 411005
Tel - (020) 25512336/37/39, Fax - (020) 25511379
Email : niralipune@pragationline.com

➢ **DISTRIBUTION CENTRES**

PUNE

Nirali Prakashan
(For orders within Pune)　:　119, Budhwar Peth, Jogeshwari Mandir Lane, Pune 411002, Maharashtra, Tel : (020) 2445 2044, Mobile : 9657703145
Email : niralilocal@pragationline.com

Nirali Prakashan
(For orders outside Pune)　:　S. No. 28/27, Dhayari, Near Asian College Pune 411041
Tel : (020) 24690204; Mobile : 9657703143
Email : bookorder@pragationline.com

MUMBAI

Nirali Prakashan　:　385, S.V.P. Road, Rasdhara Co-op. Hsg. Society Ltd., Girgaum, Mumbai 400004, Maharashtra;
Mobile : 9320129587 Tel : (022) 2385 6339 / 2386 9976, Fax : (022) 2386 9976
Email : niralimumbai@pragationline.com

➢ **DISTRIBUTION BRANCHES**

JALGAON

Nirali Prakashan　:　34, V. V. Golani Market, Navi Peth, Jalgaon 425001, Maharashtra, Tel : (0257) 222 0395, Mob : 94234 91860;
Email : niralijalgaon@pragationline.com

KOLHAPUR

Nirali Prakashan　:　New Mahadvar Road, Kedar Plaza, 1st Floor Opp. IDBI Bank, Kolhapur 416 012, Maharashtra. Mob : 9850046155;
Email : niralikolhapur@pragationline.com

NAGPUR

Nirali Prakashan　:　Above Maratha Mandir, Shop No. 3, First Floor, Rani Jhanshi Square, Sitabuldi, Nagpur 440012, Maharashtra Tel : (0712) 254 7129;
Email : niralinagpur@pragationline.com

DELHI

Nirali Prakashan　:　4593/15, Basement, Agarwal Lane, Ansari Road, Daryaganj Near Times of India Building, New Delhi 110002
Mob : 08505972553, Email : niralidelhi@pragationline.com

BENGALURU

Nirali Prakashan　:　Maitri Ground Floor, Jaya Apartments, No. 99, 6th Cross, 6th Main, Malleswaram, Bengaluru 560003, Karnataka;
Mob : 9449043034
Email: niralibangalore@pragationline.com
Other Branches : Hyderabad, Chennai

niralipune@pragationline.com　**|**　**www.pragationline.com**

Also find us on　 **www.facebook.com/niralibooks**

Preface ...

The science which deals with the earth is known as Geology (geo means 'earth'; logos means 'study or science'). It deals with all the physical and chemical aspects of the earth, along with its structure, origin, age and evolutionary history of its surface. Its study is facilitated by the information obtained from specialised branches of science, such as Physical Geology / Geomorphology / Dynamic Geology, Mineralogy, Petrology, Structural Geology, Stratigraphy, Palaeontology, Economic Geology, Geochemistry and Geophysics.

Geology is an inter-disciplinary science and the advance made in its understanding is based on valuable research contributed from the other sciences. This has given rise to a host of specialised fields such as environmental geology, marine geology, photogeology and mining geology, to name a few.

This book has been written for the first year students of Geology and also for those who have opted for Geology as one of their subjects for competitive examinations such as UPSC, MPSC and others.

This book attempts to provide a basic understanding of the subject of geology to the students of F.Y.B.Sc. of the Savitribai Phule, Pune as per the revised syllabus w.e.f. June 2019. It covers the topics such as stratigraphy, palaeontology and petrology. Along with the students from Pune, this book will also be useful for the students of Shivaji, Kavayitri Bahinabai Chaudhari North Maharashtra and Goa Universities, as well as, other Indian Universities. This book, written in a simple and lucid language and well illustrated with figures and sketches, will be of immense help to the students.

The students, therefore, won't have to spend their valuable time in searching for the topics of study in various reference books or to rely entirely on the notes made in the class. We hope that the book will fulfill the expectations of the students, teachers and all other readers.

We are grateful to all the colleagues from different Departments of Geology, especially from Nowrosjee Wadia College, Fergusson College, Poona College, Sinhgad College and Z.B. Patil College (Dhule) for the help they extended during the preparation of the book.

We are extremely thankful to Mr. Dineshbhai Furia, Mr. Jignesh Furia and all the staff of Nirali Prakashan which includes Prachi Sawant, Roshan Shaikh and Ravi Walodare for their efforts and keen interest in the completion and publication of this book.

We would appreciate and welcome any suggestions or constructive comments for the improvement of the text.

Pune   **Authors**

Syllabus ...

GL 121: Stratigraphy and Palaeontology (2 Credits)

Credit I: Principles of Stratigraphy and Palaeontology I

A) Introduction, definition, principles of stratigraphy, development of stratigraphic concepts and importance of stratigraphy (3)

B) Palaeontology: Definition, branches, importance and scope. (2)

C) Fossils: (5)

- Definition, conditions and modes of preservation of fossils.
- Techniques used in collection (Spot and channel), preservation &
- Illustration of mega fossils
- Uses and Importance of Fossils

D) Systematic position, morphology of hard parts, geological and geographical distribution of the following:

Phylum Echinodermata

Class Echinoidea: Morphology of hard parts of Regularia.

Variation in the apical disc in echinoids.

Phylum Arthropoda

Class Trilobita: Morphology of hard parts of Trilobites

Credit II: Palaeontology - II

A) Phylum Mollusca: (7)

I. Class Lamellibranchia or Bivalvia: Morphology of hard parts of the shell, ornamentation and types of hinge lines.

II. Class Gastropoda: Morphology of hard parts of the shell and forms of the gastropod shell.

III. Class Cephalopoda: Morphology of hard parts of Nautilus, Ammonoids type of suture lines and Belemnites.

Comparison between Nautilus and Ammonoids

B) Phylum Brachiopoda (4)

Morphology of hard parts of Class Articulata and Inarticulata.

Types of brachial skeleton. Comparison between Lamellibranchs and Brachiopods.

D) Phylum Coelenterata (2)

Class Anthozoa: Madreporaria, polyp, medusa, types of septa.

E) Concepts of organic evolution (Definition, Evidence of evolution, Macro & Micro evolution, Darwinism, Lamarckism & Mutation) (2)

GL 122: Petrology (2 Credits)

Credit I: Introduction to Petrology and Igneous Petrology

Petrology

A) Definition and major divisions (3)

    a. Definition of petrology, lithology, petrography, petrogenesis.

    b. Major divisions and diagnostic characteristic of rocks: igneous, Sedimentary and metamorphic.

    c. Rock cycle.

Igneous Petrology

B) Magma (2)

    a. Magma and its composition.

    b. Bowen's reaction series (sequential crystallization of minerals in the igneous rocks)

    c. Formation of crystals and glass.

C) Forms of Igneous bodies (3)

    a. Intrusive: Concordant and discordant intrusions

        1. Concordant: sill, laccolith, lopolith.

        2. Discordant: dyke and veins, batholith.

    b. Extrusive: lava flows

D) Textures and Structures (4)

    a. Textures: Definition and factors controlling textures: Equigranular (granitic), Inequigranular (porphyritic), glassy

    b. Structures: Vesicular, amygdaloidal, blocky, pillow, flow and columnar.

E) Classification of Igneous Rocks (3)

    a. Basis of Classification: Depth of formation, silica percentage, type of feldspar content and colour index.

    b. Petrographic description of Igneous rocks and Tabular classification

Credit II: Sedimentary Petrology and Metamorphic Petrology

Sedimentary Petrology

A) Sediments, sedimentation, sedimentary environment (definition and types) and formation of sedimentary rocks – Weathering (mechanical and chemical), erosion, denudation, transportation, deposition, compaction, cementation and lithification Classification and description of sedimentary rocks **(5)**

B) Textures and Primary structures of sedimentary rocks: **(3)**

   a.   Clastic and non-clastic textures.

   b.   Structures: Lamination, bedding (concordant and discordant), graded bedding, ripple marks and mud cracks/sun cracks.

Metamorphic Petrology **(4)**

A) Definition of metamorphism, agents of metamorphism, kinds of metamorphism, characteristics of different types of metamorphism **(4)**

B) Structures in metamorphic rocks: maculose, slaty cleavage, granulose, schistose and gneissose **(1)**

C) Tabular classification giving original rock, predominant agent of metamorphism, type of metamorphism and their metamorphic product of the following rocks: Slate, Quartzite, Marble, Hornblende schist, Mica schist, Hornblende gneiss. **(2)**

Contents ...

❖ ❖ ❖

Chapter 1 ...

INTRODUCTION TO STRATIGRAPHY

Contents ...

1.1 INTRODUCTION

The study of 'layered rocks' is called stratigraphy. This term has been derived from the word *stratum* meaning 'layers' (usually sedimentary rocks) and *graphy* means 'description'. It deals with the geological history of the earth's stratified rocks with regard to the time of their formation. Thus, stratigraphy does not restrict itself to a descriptive study of sedimentary rocks, but also includes their formation, distribution, age and correlation with other rocks. It can be considered as the repository of the past geological record.

1.2 DEFINITION

Stratigraphy is a branch of geology that deals with the study of rock layers and layering. As most of the layered rocks are deposited as sediments in water, the term 'stratigraphy' is primarily used in the study of sedimentary rocks, although some layered rocks are volcanic in origin. As stratigraphy covers a vast spectrum of the earth's geological history, it has been linked with various branches in geology. Earlier, stratigraphy was defined by geologists on the basis of their own fields of specialization, e.g., a palaeontologist defines stratigraphy as that 'branch of geology which deals with the occurrence, distribution, chronology and age of the rocks, as based on the fossils present in these rocks'.

A palaeo-geographer defines stratigraphy as 'a science which helps to study the geography of the earth in the past ages'. A geochemist or a geophysicist defines stratigraphy in a manner which reflects the geochemical or geophysical evolution of the earth. Krumbein and Sloss consider stratigraphy as 'an integrating discipline which combines data from all other branches of the earth sciences so that a definite order of historical geology emerges'. Carl Dunbar defines stratigraphy in its restricted sense and has subdivided the branch of stratigraphy into three stages namely, 1) description of the stratified rocks, 2) their correlation and 3) determination of their mutual time relation.

Weller defines stratigraphy as 'a branch of geology which deals with the study and interpretation of stratified and sedimentary rocks, their identification, description, sequence (both vertical and horizontal), mapping and correlation of stratigraphic rock units'.

Taking into account all these definitions it is clear that the study of stratigraphy involves information gathered from several branches of geology. This information is then arranged in such a way that a systematic historical account of the earth is presented. Thus, stratigraphy is linked with all branches of geology and also to most other sciences. Thus, any geological process which was operative in the geological past or is in operation today, is well within the scope of stratigraphy. This is best illustrated by the changing patterns of land and sea, fluctuating climates and also the evolution of life on the earth.

Stratigraphy is the basis for understanding of the geological history both, locally and even on a global scale. It helps in arranging the geological history of rocks in order (geochronology of geological events). These studies are also useful in understanding the palaeoclimatic changes. The correlation of different rock formations, exposed in different continents, is possible due to stratigraphic studies. The understanding of continental plates, their disposition in the geological past and their evolution (plate tectonics), has been made easier by stratigraphic studies.

Geologists over the years, have recorded their observations with regard to stratified rocks on the earth's surface. However, it was only in the 17th century that the 'time factor' was considered to be an important aspect in unfolding the earth's history. Based on rational scientific data, it was proved that stratified rocks have developed over a period of millions of years. Initially, the superposition of strata was noted and it was also realized that after the deposition of a particular layer or bed there was a time lapse or interval of time before the next bed was deposited. This time gap in deposition which is an important factor was noted. The entire stratigraphy is thus based on these three principles.

1.3 PRINCIPLES OF STRATIGRAPHY

There are three principles in the study of stratigraphy, namely

1) Uniformitarianism,

2) Order of superposition and

3) Faunal succession / palaeontological correlation.

1) Uniformitarianism:

The idea of uniformitarianism was presented by the Scottish naturalist James Hutton. This principle states that the geological processes which have acted in the geological past were essentially the same as those in operation today. It is best summarized in the statement 'the present is the key to the past'.

Today, as we get to see the sedimentary layering and variation in the grain size of the sediments along the river banks, is due to the fluctuations in the velocity of the river. Similar structures and textures are seen in the sedimentary rocks that were formed in the geological past. According to the principle of uniformitarianism, these rock structures and textures have resulted from similar processes. Sedimentary features like ripple marks, graded and cross bedding are all a result of transport of matter as it occurs even today and so are the effects of other geological processes.

2) Order of Superposition:

This is another important principle put forward by the Danish physician, Sir Steno, in 1669. He pointed out that in an undisturbed sedimentary sequence, the order in which the layers are laid down is starting from the bottom and continues to the top. This means that the bottom layers are older than the layers above it (and is also applicable to the deposits within a single layer). Such layers, which are known as laminae, are indicative of the relative age of the rocks even if they are developed in different parts of the same region. However, when the stratified rocks are folded, overthrusted or faulted, it is first necessary to determine whether the rocks are in their original position or whether they have been overthrusted or overturned. This can be determined by noting the sedimentary structures in the rocks, such as cross bedding, current bedding and ripple marks. From the principle of order of superposition the following two conclusions can be drawn:

i)   Igneous rocks are younger than the rocks into which they intrude (Principle of cross-cutting relationships).

ii) A fault is younger than the formation it cuts.

3) Faunal Succession / Palaeontological Correlation:

A British surveyor, William Smith who is known as the 'Father of Modern Stratigraphy', was the first to realize that fossils can be used to correlate the strata in which they are found. During the course of his work he realised that the same group of fossils always occurred in the same rock layers. This similarity in fossils also indicated their similarity in age. Fossil species in the layers below and above a certain bed were distinctly different from those in that bed. Also the fossils occurred in the same order in the rocks in widely separated areas. This observation by William Smith led to the establishment of the principle of faunal succession. It explains that the varied fossil assemblages associated with different sedimentary formations are the result of organic evolution. Thus, the primitive fossils occur in the older rocks whereas the fossils of the more evolved organisms are to be found in the younger rocks.

1.4 DEVELOPMENT OF STRATIGRAPHIC CONCEPTS

The development of the stratigraphic column is the most significant achievement in geology. Its development started during the end of the 18th century. Prior to the advent of stratigraphy, the earth was generally considered to be significantly younger than a million years. An estimate by Irish Archbishop James Ussher in the 17th century, showed that the earth was younger than 6000 years. Ussher's estimate was based on his chronology of the Old Testament. Such misconcepts were later on rejected when systematic geological studies were done. Subsequently, most of the divisions were developed over a period of 50 years in the first part of the 19th century to develop the modern version of the stratigraphic concept (Fig. 1.1).

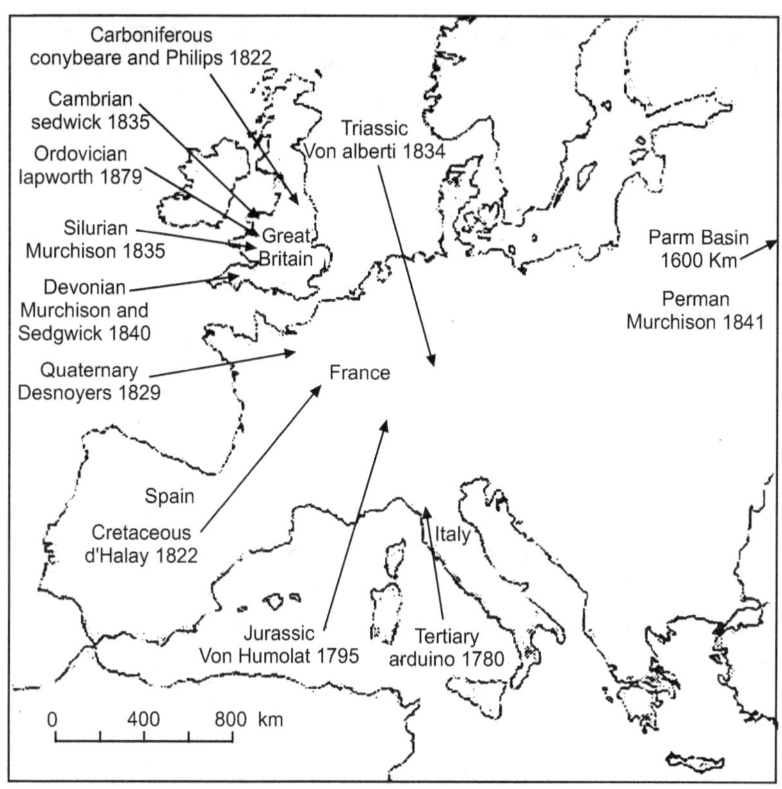

Fig. 1.1: Locations of the earliest studies of stratigraphic column

As shown in the Fig. 1.1, most of the systems were established from the study of the stratigraphic record in Europe and were initially defined on the basis of lithology or rock types alone. For example, the Triassic was divided into three parts by Von Alberti in Germany in 1834. This stratigraphic column was further modified by Murchison in 1835 when he defined Silurian system using fossils as an evidence. Later on, this helped Darwin while proposing his ideas about organic evolution. However, it must be remembered that William Smith, the canal surveyor, had already published his geological map of Britain in 1815 and he had adopted a holistic approach based on all available observations.

Stratigraphy was developed in England during the early 1800s with the work of a land surveyor named William Smith. Using stratigraphic concepts, 19th century geologists demonstrated that the earth was far older than a million years. For example, sedimentation rates of marine limestone typically range from about 2 to 20 cm per thousand years. At these rates, about 6000 m of marine limestone exposed in southern Nevada required at least 30 to 300 million years of continuous deposition. Additionally, these rocks overlie rock units that show a complex history of metamorphism, intrusion, faulting, and uplift prior to deposition of the limestone. All these observations added more 'scientific' data to the previous observations. Some of these older concepts were then either rejected or were modified.

The stratigraphic column is now universally accepted across the world and the names of the individual time units have been standardized. The one exception to this is the Carboniferous which was first named in Britain after the vast coal deposits of that region. However, only the top half of this period contains coal. The bottom part is composed of marine limestones and in recognition of this, the USGS in 1953 further subdivided the system into Mississippian and Pennsylvanian. Hence, this terminology is only widespread in America and not in Europe.

The chronostratigraphic scale is a summation of all stratigraphical knowledge and as such there is no one cliff or quarry section on the Earth at which all units are exposed. It forms the basis of all geological maps and correlation. As it forms such an important part of geological study, the development of the stratigraphic column through the eighteenth and nineteenth centuries follows the thinking and expansion of the geological perspective on the age of the earth.

1.5 IMPORTANCE OF STRATIGRAPHY

It is one of the major branches of geology which provides the basic information with regard to stratified rocks. Stratigraphy lies at the core of any geological study as no investigation of earth's history can be completed without better understanding of vertical stratigraphic column. It provides means of joining together the world's geologic records. It is most useful when an area is being prospected for any economically important mineral or for oil. The strata suitable for their accumulation in the subsurface are determined on the basis of their stratigraphy. Stratified rocks also contain almost all the fuel deposits such as coal and petroleum; radioactive minerals, iron ores, phosphates, salt, gypsum and limestone, along with huge reservoirs of ground water, to name only a few.

POINTS TO REMEMBER

- The study of 'layered rocks' is called stratigraphy.
- The word *stratum* means 'layers' and *graphy* means 'description'.
- The term stratigraphy is primarily used in the study of sedimentary rocks.
- A palaeontologist defines stratigraphy as that 'branch of geology which deals with the occurrence, distribution, chronology and age of the rocks, as based on the fossils present in these rocks'.
- Krumbein and Sloss define stratigraphy as 'an integrating discipline which combines data from all other branches of the earth sciences so that a definite order of historical geology emerges'.
- There are three principles in the study of stratigraphy, namely:
- Uniformitarianism: The geological processes which have acted in the geological past were essentially the same as those in operation today ('the present is the key to the past').
- Order of superposition: In an undisturbed sedimentary sequence, the order in which the layers are laid down is starting from the bottom and continues to the top.
- Faunal succession: The primitive fossils occur in the older rocks whereas the fossils of the more evolved organisms are to be found in the younger rocks.
- The chronostratigraphic scale is a summation of all stratigraphical information.
- Investigation of earth's history can be completed only with understanding of vertical stratigraphic column.

EXERCISE

1. Define stratigraphy and explain various ways of defining the stratigraphy.
2. Explain in detail the development of stratigraphic concepts.
3. Write notes on:
   a) Importance of stratigraphy
   b) Principles of stratigraphy
   c) Development of stratigraphic concepts
   d) Importance of stratigraphy

Chapter **2**...

INTRODUCTION TO PALAEONTOLOGY

Contents ...

2.1 INTRODUCTION

The term palaentology has been coined from the Greek terminology (*Palaeo-* ancient, *onto*-life and *logos*-study). The subject of palaeontology involves the study of innumerable dead remains of microscopic and megascopoic animals and plants that once lived on the earth's surface and are now a part of the stratified rocks. As it encompasses the study of organisms and plants encountered from within the rock masses, its fundamentals are derived from the two main science streams namely, biology and geology.

Definition :

Palaeontology can be defined as 'the study of fossils as a guide to the history of life on Earth'.

It may be considered as the study of life of the palaeo-geologic time, based on the preserved dead remains or indications of plant and animal assemblages which may or may not have any resemblance with the living species. The earliest studies on palaeontology date back to 576-486 B.C., when Xenophanes examined and identified marine shells on a hilltop and interpreted that the area must have been under the oceanic waters in the geological past. This subject was not given much importance until the 17th century when Danish physician **Nicolaus Steno** (1638-1686) carefully observed the fish tooth in the sediments and identified its similarity with that of the living sharks. Prior to his time, the fossils were considered as the fantasies or sport of nature or devil's work, which included skeptical observations and interpretations based on the then available scientific information. Fundamentals of this subject were probably brought into practice by **William Smith** (1769-1839) whose observations on the sediments and their characteristic fossil assemblages led to the Law of faunal succession. Subsequently, **Cuvier**, on the basis of varied fossil contents in the sediments, identified the presence of catastrophic events which caused the death of flora and fauna (mass mortality). Cuvier also noted that the younger sediments that overlaid older ones had modified fossil assemblages. **Chevalier de Lamarck** has been honoured as the *"Father of Invertebrate Palaeontology"* because of his dedicated work on organic evolution. Later on, **Charles Darwin** went through his investigations and its modified version was proposed as the 'Theory of organic evolution'. It is a cumulative effect of all such vital observations that the subject of Palaeontology has gained the status of a special branch of geology.

The fundamental units of these studies are obviously the fossils. British Palaeontologist, **Henry Woods**, defines fossils as the 'remains of animals and plants of past ages, preserved in rocks'. The term fossil has been coined from Latin word *fossilis* meaning to dig up. Prior to the birth of the science of Palaeontology, the dead remains were not treated as visible evidences of the life that might have existed in the geological past. The fossils are of mainly two types, body fossils and trace fossils. The body fossils are directly preserved parts of the original plants or animal remains. It is the process of burial of the dead organism or plant in the sediments that are in the process of formation. Further, these dead remains are slowly converted into rocks by a process called petrification.

With the advancement of scientific knowledge, not only the dead remains of the plants and animals but even their traces have now been enrolled as fossils. In living conditions, the organisms are constantly moving on and underneath the surfaces of the sediments. For example, crab moves along the surface of the sea beach or burrows itself for shelter. Such actions by the organisms for their feeding, shelter and movement create tracks, trails and borings on or within the sediments. Subsequently, the burial of these sedimentary horizons preserves such traces as trace fossils. Therefore, the term fossil is assigned to any trace of the past life. Hence, along with the actual remains of organisms and plants (e.g., bones, shells, leaves, tree trunks, seeds, etc.) even the effects of their activity preserved in *situ* within the rocks (e.g., burrows, footprints, organic compounds produced by organisms, etc.) are also called as fossils. *Pseudo fossil* is the term associated with deceptive structures and shapes that are created by the nature, e.g., the passage of iron-rich or manganese-rich solutions through the joint or bedding planes creates a branching tree-like pattern resembling fossils.

2.2 BRANCHES OF PALAEONTOLOGY

Due to the advancement and specialisation in the scientific world, there was a need for the further research with regard to the different aspects of fossil study. Few of the specialised branches of palaeontology are as follows:

(a) Palaeobiology: This branch of palaeontology deals with the study of plant and animal remains in the geological sediments of the past. It involves the understanding of the modes of habitat, evolution and extinction of plants and animals. It is further subdivided into two branches namely, palaeozoology and palaeobotany. Former is the study of organisms that have been broadly classified as vertebrates and invertebrates. Vertebrate palaeontology deals with the animals with backbone (like fishes, reptiles, birds and mammals), whereas invertebrate palaeontology deals with animals which do not posses a backbone, e.g., molluscs, corals and protozoans. The science of palaeobotany deals with the study of fossil plants and to obtain the record of the geological history of the plant kingdom.

(b) Micropalaeontology: The study of microfossils constitutes the science of micropalaeontology. The microscopic studies allow understanding of finer details of the fossils which can not be identified with the help of naked eyes, e.g., foraminifers are sensitive to the environment in which they live and hence, their studies allow us to visualise the palaeo-environments of the past. Recent development in the field of micropalaeontology has given rise to a new a branch dealing with the pollens and spores of ancient plants. This branch, therefore, has been granted the status of a separate science stream called as palaeopalynology. Sometimes, microscopic fragments of animals get preserved within the sediments and their study under microscope enhances the minute morphological details.

(c)   Ichnology: It is a branch of palaeontology that deals with the study of tracks, trails and burrowing habits of organisms, e.g. footprints of dinosaurs and work-trails.

2.3 IMPORTANCE OF PALAEONTOLOGY

As the subject of paleontology explores prehistoric life, using fossils as tools, it is possible to predict about the organism evolution, as well as, interactions of life forms with one another and their environments. Today, palaeontology has become an interdisplinary science to unfold the earth's ancient environment that uses biological, archaeological and geological evidences, in association with computer and remote sensing techniques. The engineering and biochemical techniques are also made use of in the palaeontological research. The use of mathematical models has become necessary in speculating ancient life forms and their environments (e.g., terrestrial, aerial and marine), and so on.

Scope of Palaeontology :

The students of palaeontology (geology) are gaining more scope all over the globe because this subject is useful in providing help in many aspects of human development. As ancient environments can be explored by palaeontological research, its applicability for predicting of environments of the present and future have become easier. These studies have also enhanced the exploration for oil, minerals and other important natural resources. The usage of computer systems, mathematical modelling and remote sensing has become essential for dating and correlating geological events of the past. The digital mapping and data analysis skills are vital for interpreting geological events on global scale. Hence, a palaeontologist, with advanced skills and technology, is being preferred with priority. There are numerous geological and archaeological research institutes in India and abroad which are employing qualified paleontologists. A palaeontologist can also take up teaching and research assignments in geology and allied subjects in the colleges, universities, Geological Survey of India, Palaeontological Research Institution, oil and natural gas sectors and so on.

2.4 FOSSILS

Following are the prerequisites for any substance to be called as a fossil:

1. The fossil has to be an antiquity, the literary meaning being 'quality of being very old'. It means that any substance can be called as a fossil only when it is a part of any organism or plant that existed in the geological past. However, there is no fixed age limit for the dead remains to be considered as fossils. It is agreed upon amongst palaeontologists that the dead remains of pre-historic period or prior to that should be called as fossils. For this reason, the dead remains of the historic period are not fossils.

2. The fossils should be organic in origin, i.e., they should be formed after the death of living fauna and flora. But, not all organically originated organisms can be considered as fossils. For e.g., Petroleum, that has been organically originated, cannot be called as a fossil. The trace fossils are also exceptions to this rule. The formation of trace fossils is governed by the activities of the organisms but their organic remains are not preserved as body fossils. Still they are regarded as fossils.

3. It is usually expected that the fossils would not have any place in the present biosphere. However, there are some animals which evolved millions of years back and are still surviving in the present environment. For e.g. the animal like Nautilus which evolved about 550 million years ago is still surviving by its four species in the modern oceans. Hence, such animals are called as living fossils.

4. The main characteristic of any fossil is its preservation in the sediments. But, exceptional occurrence of insects in amber or animals in frozen ice can be cited as examples of fossils that have not undergone burial in the sediments.

5. The process of fossilization is now referred to as a branch of palaeontology called **Taphonomy**.

2.5 CONDITIONS NECESSARY FOR FOSSILIZATION

It is not necessary that all plants and animals of terrestrial or aquatic environment have to be preserved as fossils. In fact, the remains of most of the organisms are easily destroyed due to their decay and erosive action of the exogeneous and endogeneous processes of nature. As a consequence of which it finds no chance of being preserved in the sediments.

The dead organisms can be categorised into two main types namely, organisms with hard parts (skeleton) and the ones without hard parts. Animals with hard and rigid skeletons stand much better chance of being converted into fossils. Hence, the plants and animals which do not possess bony skeletons are readily decomposed under terrestrial or aquatic environments. For example, the leaves, buds, flowers and fruits along with animals like jellyfish, insects and worms (which are devoid of any hard parts), leave no record of their existence. It is only under rare conditions that such remnants are preserved as imprints in the sediments.

Although, the hard parts of organisms are quite strong enough to withstand wear and tear for some length of time, they do get disintegrated and destructed due to their prolonged exposure to weathering agents on the surface. But, the same organism or plant may be preserved if it undergoes rapid burial under the sediments. Thus, the terrestrial animals or plants have rare opportunity for their fossilization. Aquatic environments allow deposition of uniform sedimentation which readily covers up the remains of the plants and animals. Hence, the fossilisation finds a much better chance in aquatic environment as compared to the terrestrial environment.

The necessary condition favourable for a better preservation includes a) Organism must possess some hard parts or skeleton and b) their remains should be covered up by a thick pile of sediments.

The necessary condition for fossilization as mentioned above are no doubt important but may not be sufficient for the conversion of the remains into well formed fossils.

2.6 MODES OF PRESERVATION OF FOSSILS

The fossil record is observed in many forms, each being dependent upon the characteristics of the organism, the media in which it was embedded and the chemical actions that the organism has undergone after its burial.

As discussed earlier, the organisms leave behind two major types of fossil records namely, the skeletal remains of the organisms as body fossils and the other in the form of trace fossil. These two types can be preserved as either unaltered or altered remains within the sediments.

(1) Unaltered remains: It includes the fossils whose hard and soft parts are preserved in toto, i.e., body parts remain intact with little or no alteration from its living state. Preservation of unaltered remains is possible in two ways.

(a) Entire organism preserved intact: It includes intact preservation of soft and hard parts of the organisms. For example, extinct woolly mammoth and rhinoceros of Siberia have been preserved under the natural refrigeration of thick ice cover for many thousands of years. Their bodies have suffered negligible alteration since burial so that even after 45,000 years their eyes, skin, blood, flesh and partly digested food have remained without much degradability. Along the Baltic coast, the insects and spiders have been engulfed in the amber (tree resin) and preserved without any physical or chemical alterations.

(b) Partially preserved skeletons: In this type of preservation, the soft and delicate parts of the body are preserved occasionally but the hard skeleton is preserved almost unaltered. Fossils of this type exhibit perfect preservation of the original composition and structure of hard and durable parts of the skeletons. An example of this type is that of Santana formation of North Brazil where three dimensional fish fossils have been preserved in cretaceous limestones.

(2) Altered remains: Due to chemical changes that occur within the sediments, the preserved fossils are subjected to minor or major chemical alterations such as leaching. This is facilitated by the ground waters which are enriched with carbonic acids that are able to dissolve most of the aragonitic skeleton as bicarbonate of lime.

(a) Leaching: Most of the fossils undergo leaching effects subsequent to burial. The main reason for the fossils to undergo leaching is their calcareous composition. Most of the organic remains are made up of calcium carbonate (calcite and aragonite), of which, calcite is more stable than the aragonite. Other fossils may be formed of calcium phosphate, chitin, silica and scleroprotein. When such fossils are exposed to the circulating ground waters, they get partially or completely dissolved. This action is aggravated in the coarser sediments (which facilitate the movement of solutions) and in the presence of acidic or alkaline conditions. Due to leaching, the fossils are corroded along their exposed parts resulting into pitted, roughened surfaces, dullness of the lustre along with loss of original weight. Typical examples of this type are molluscan shells of Cenozoic era whose outer covers (periostracum) have been completely leached away.

(b) Permineralisation: Due to the chemical susceptibility of the shells and skeletal parts to the solutions percolating through the sediments, the fossils are commonly altered by the addition of certain inorganic substances. Calcium carbonate ($CaCO_3$) and silica (SiO_2) are commonly enriched into the interstitial parts of the fossils through circulating ground waters.

Due to these inorganic ingredients, the fossils gain weight and often swell to become massive. But, due to Permineralisation, the fossils are able to resist further alteration and erosion.

(c) Petrification: The petrification implies the conversion of plant remains into a rock-like forms in which the organic matter of the fossils gets replaced by the inorganic minerals. In this process, the removal of organic matter is simultaneously replaced by precipitation of an equal quantity of replacing minerals such as silica (SiO_2), calcium carbonate ($CaCO_3$) and oxides / sulphides of iron. It is a molecule by molecule replacement of one substance by the other. As it proceeds very slowly, the replaced plant cells or tissues retain their original shapes and even most delicate structures of the plants (even the growth rings can be traced). At the end of this replacement, the plant gets modified into a hard rock-like matter. The process of petrification is best exemplified by the tree trunks of the Yellow Stone National park (United States), Cretaceous rocks in Ariyalur (South India) and Jaisalmer (Rajasthan).

(d) Carbonisation or distillation: During carbonisation, the organisms are decomposed so as to loose their nitrogen, oxygen and volatile contents after having buried in the sediments. As a result, they become relatively enriched with the carbon contents and hence, the term carbonification. The coal seams all over the world are the best examples of carbonized plant remains. During carbonisation, these fossils are also subjected to compression. The fossils that have undergone carbonisation include fauna of coal seams at Raniganj, Jharia in West Bengal. Due to the elimination of oxygen, nitrogen and other volatile constituents in carbonization, it is also called as distillation.

(e) Moulds and Casts: The most common chemical change associated with buried fossils is the solution effect. When the shells are completely embedded in the sediments, the soft parts of the animal get decomposed leaving behind an empty cavity which subsequently gets filled in by the sediments. These sediments are, later on, converted into compact and hard rocks. The percolating dilute acids that enrich the ground waters react with the embedded shell matter and consequently bring about leaching effects. This leaves behind cavities proportionate to the size and shape of the fossils (Fig. 2.1).

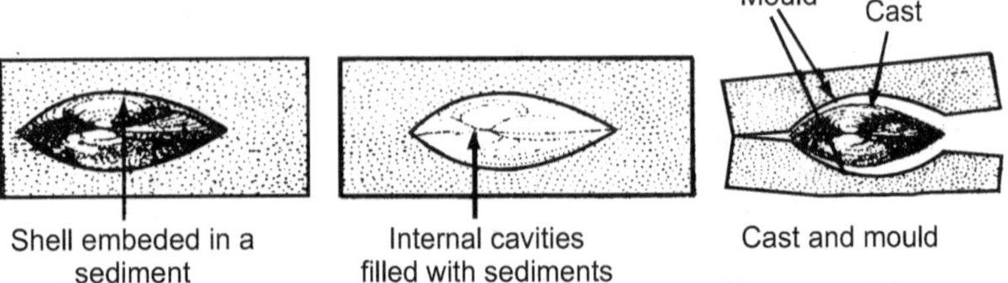

Fig. 2.1: Moulds and casts

These hollows with the shapes of the original fossils preserve the external characteristics of the shells such as ornamentation, growth lines and crenulations. These cavities which possess most of the external characteristic features of the shell are called external moulds.

In case of some moulds, the sediments which have filled the internal cavity of the shell retain the replica of internal features of the shell (Fig. 2.1). These are termed as internal moulds. For educational purpose, artificial external moulds are often produced by pressing the external surface of the fossil into the soft Plaster of Paris and then its hardened statue-like portion is used as an external mould for the study.

The cast is formed when the interior of the shell is filled with sediments or minerals of various types and later on, the original shell gives way to the dissolution (Fig. 2.1). When the shell dissolves away, the remaining sediments are left with the pseudomorphs of the original shells, called as casts.

(f) Imprints: Most of the plants and animals, devoid of their hard parts, leave behind a record of their existence in the fine grained sedimentary rocks like clays and mud stones. These are imprinted when the clays and other argillaceous rocks are in their unconsolidated states. When these sediments get compacted and cemented to form hard rocks, they form the imprints. Due to the subsequent erosion of these rocks, the imprints are exposed on the surface, e.g., impressions of jelly fish and leaves in clays and shales (2.2).

Fig. 2.2: Impression of a leaf on fine-grained sediment

(g) Tracks and trails: As stated earlier, along with the actual remains of the organisms, their traces are also considered as fossils. Vertebrates and invertebrates both leave behind the record of their existence on the soft sediments. Of these, tracks, trails and footprints depict the traces of locomotory organs on the surface of the sediments and on the sea bottom. For example, the tracks of the reptiles in the Upper Triassic are well known as they indicate not only the type of organism but one can also predict about their shapes, sizes, lengths of the limbs and posture of the animal. The trails are the impressions of the movement of the organism. For example, the worms produce trails during their movement in search of food and shelter. Different trace fossils have been illustrated in the Fig. 2.3.

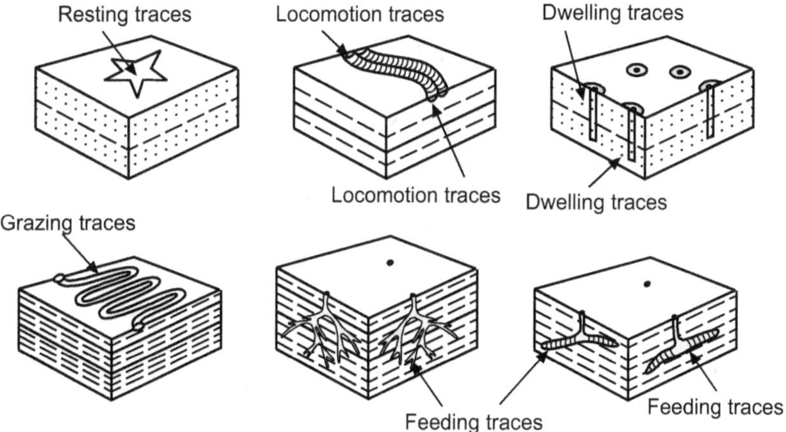

Fig. 2.3: Types of trace fossils

Usually, the tracks are produced by molluscs, burrows are formed by worms and the sponges create borings. The specialised branch of palaeontology which deals with the study of trace fossils is called as Ichnology.

(h) Coprolites: These are the fossil faecal pellets of the pre-existing organisms. This study of fossil excreta may provide valuable information pertaining to the food habits of the organism and hence, based on the size of the faecal matter, the type of fossil species can be recognised.

2.7 COLLECTION AND PREPARATION OF THE FOSSILS

To have a meaningful data in palaeontology, it is essential that one must have a good collection of fossil contents which are representative of the age-wise geological events of the past. In the palaeontological studies, key to the success lies in the skill of selecting proper sites for fossil collection. It is very important to know where and how to look for fossils and to use the most effective method for collecting fossils. The field work for fossil collection is initiated when the field equipments are chosen to suit the collection of fossils.

Selection of field sites and equipments: The first and foremost thing a palaeontologist must know is 'where to look for the fossils?' The best way towards this is to accompany an experienced palaeontologist who has had an earlier field experience in the same terrain. If this is not possible then the published research papers on the already surveyed localities can be the best guides. In a virgin terrain, however, one must rely on extensive field work with the help of topographic sheets. In addition, some of the thumb rules also may help in selecting the collection sites. For example, the igneous and metamorphic rocks are of no use for the preservation of fossils due to their high temperature and pressure conditions. The sedimentary rocks, especially the marine sediments, can be easily looked for if they occur as relatively flat horizons with extensive dimensions. But care has to be taken that these have not been subjected to either folding or any other physical and chemical deformations. Similarly, the sediments laid down in the shallow sea, palaeo-marine sediments, deltas, extensive lakes etc. can also be expected to be fossiliferous. The sediments of continental or terrestrial origin may or may not prove to be the ideal as they are likely to have undergone erosive action.

The places like quarries, highway cuts, valleys, gullies, stream beds, sea beaches, etc. may expose fossils that have stood the test of nature's erosive mechanism. The quarries and roadside cliffs may possess rich fossil contents if their surfaces are fresh and unweathered. Under adverse conditions, exceptional occurrence of fossils may be encountered when freshly ploughed fields expose hidden fossils from the fossiliferous horizons underneath. The working coal mines, if any, also can be hunted for large number of fossils. Even abandoned coal mines may prove fruitful if surveyed along their dumping pits. The use of hand lens is very useful in locating the exact position of the fossil.

The instruments needed for the fossil collection include a well-bound pocket diary, topographic maps, fossil guides, clinometer compass, geological hammer, a pair of chisels, wrapping material, paper or cloth bags, small paper labels, magnifying lens, stiff bristle point brush of whisk broom, sledge hammer or shovel, camera, marking pens with water proof ink, etc. The diary for a paleontologist is a must for sketching the fossils and for taking down the geological and geographical descriptions. The topographic maps and clinometer compass together help the paleontologist to locate the desired location and in the preparation of the geological map of the terrain. Few of the published fossil-guides or the research papers may be vital to guide the paleontologist to the desired location.

It is always advisable to collect the fossils in situ and the fossils that have been transported from their original locations (stray fossils) through streams or rolled down along slopes should be avoided. In case of weathered rocks or rocks that have undergone extensive erosion, there is a possibility that fossils have been freed by the running water and other geological agents. Such fossil may simply be picked up, numbered, labeled and placed in the collecting bags. But, in case of compact and unweathered rocks, the use of chisels, sledge hammer, stiff bristle point brush, etc., becomes unavoidable. In case of very large fossils, it is a customary to cover the fossil with plaster of Paris to avoid its breakage during transport.

Sampling: Collection of fossils requires an intensive fieldwork with precautions of picking unweathered and uncontaminated samples. This also requires thorough referencing of preview of any published literature with regard to fossil study. The field equipments required by a palaeontologist include, geologist's hammer, chisels, brushes, polyethylene bags, cardboard boxes, field diary, marker pens, tags, measuring tapes, magnifying lens, camera and clinometer compass.

The samples of rocks containing fossils may be collected from the exposed outcrops in the area under investigation. While collecting the sediment samples from the outcrops, one must ensure to remove the weathered part before obtaining a fresh sample. Two principle types of sampling methods are a) Spot sampling and b) Channel sampling.

In **spot sampling**, the sediments to be collected are of predetermined stratigraphic levels, while **channel sampling** is mostly undertaken for large stratigraphic sequences with variation of lithology and vertical intervals ranging from 1 to 3 mts. The fossils must be collected in polyethylene sample bags to avoid any contamination. The sample bag must be sealed, labelled and important observations related to the stratigraphic outcrop must be written in the field diary. While sampling the outcrops, photographs should be taken to record certain characteristic features.

In the laboratories, the fossils are separated from the rock matrix in two ways viz. mechanical method and chemical method. The mechanical methods include physical cleaning of the fossils. After the specimens have been unwrapped, the fossils should be checked for their proper labels. Before the fossil is cleaned, it should be placed in water overnight so as to loosen the extra matrix adhering to the fossil. If the fossils are numbered with normal pens that we often use, the ink will get dissolved. Hence, the numbering of the fossils has to be done by a water-proof ink and an added care has to be taken by placing few extra labels above or below the wrapped paper. The fossils are then cleaned with the help of stiff-bristled brush or toothbrush. A wire brush may be needed for elongated and hollow specimen. Long needles and tweezers may also be used to clean more delicate fossils. Excess matrix can be cautiously removed by light hammers and chisels. The broken surfaces of the rock matrix attached to the fossils can be smoothened with a wire brush, steel wool or sand paper. In mechanical cleaning, there is a possibility that the fossil may break into pieces. These broken pieces should be properly and carefully fitted together and then placed in the trays.

In the chemical method, the use of mild acids becomes essential when the adhering rock matrix is stiff enough to be removed mechanically. In this method, only the excessive rock matrix is dissolved in suitable acid without affecting the fossil. This is done after careful detection of the chemical composition of the fossil and the rock matrix. For example, silicified wood (SiO_2) can be separated from the calcareous rock matrix ($CaCO_3$). These broken pieces should be properly and carefully fitted together and then placed in the trays.

Along with the megafossils, the fossiliferous rocks may also possess microfossils. Hence, it is advisable to bring the surrounding portion of the sedimentary rock along with the fossils. These rocks are then checked for the presence of microfossils. For this purpose, the rock is crushed after taking care not to harm the microfossils or thin slides of the rock specimens are prepared and viewed under microscope. The microfossils in the loose and dry sediments can be examined through a binocular microscope and picked up with the help of a slightly moistened camel hair paintbrush. Each microfossil can then be mounted on a special micropalaeontological slides and numbered accordingly.

Illustration of fossils: The fossils are studied with reference to their classification (like phylum, class, orders etc.) and for this reason; it becomes conventional to write appropriate and precise description along with the illustration. The illustrations are of two types namely, the drawings and photographs. The drawings of the fossils should be done carefully at the hands of an artist. This will take care of every minute detail of the fossil being enhanced. Photographic illustrations, however, give better and exact idea about the nature of the fossil. In case of some fossils having same contrast as that of its matrix, extra colours can be added to the matrix or else the photograph can be touched upon with artificial photographic colours. The fossil is photographed along different angles so as to cover all its morphological details such as aperture, suture lines and ornamentation. The microscopic studies under scanning electron microscopes (SEM) are specially recommended for the microfossils, which do not exhibit all their important features even under binocular microscope.

2.8 USES AND IMPORTANCE OF FOSSILS

Fossils are the connecting links between the past life and the present living species. Hence, depending upon the nature of their occurrence, type and the information they carry with their presence, fossils have served as guiding tools for the reconstruction of the vertical geological column in time and space. Moreover, the presence and absence of the fossils also is an indication of some typical geological events. Thus, the importance of a fossil lies in its usage and applicability to unfold the mystery of the past in different branches of geology. Following are various uses of fossils:

1. In Geology: We have learnt earlier that the science of geology constitutes the study of the earth. It deals with the origin of different kinds of rocks, their relative ages and relationship they bear with one another. This unfolding of the earth's history requires some tools such as fossils which can be taken as markers of certain climates, geological environments and water-depth conditions. The fossils have also provided important supportive evidence in correlating rocks of the similar origin all over the world. Few of these examples will illustrate the use of fossils in geological interpretation.

(a) The fossils are representatives of certain geological environments. This factor has been utilised in proving that the oldest landmass on the earth was a single piece called as *Pangea* which later on, split into two parts namely, *Gondwana* and *Laurasia*. The nature and similarities of these mega-continents have been established with the help of similar fossil contents. For example, the nature of single mega-continent was made evident by the presence of fresh-water reptile (*Mesosauras*) on the continents of South America and South Africa, which are separated by oceanic waters today. This correlation has also been proven by the similarities between the coal seams of India, Australia, Antarctica and eastern part of South America during Permo-Carboniferous period.

(b) The presence of fish and ammonoid fossils, (indicative of marine environment) on the mountainous regions of Himalaya has indicated that there existed an ocean during the geological past, called the Tethys.

(c) The presence of marine fossil bearing beds near Dwarka (Gujarat) has supported the multiple submergence of the region under the sea in the geological past.

(d) The extensive and thick pile of the Deccan Traps is a result of volcanic eruptions during the Cretaceous-Eocene period. There have been periods of quiescence in between the eruption of successive flows. During this period, the surface was exposed to weathering creating depressions (basins) in which sedimentation took place. The exact nature of the environments existing then can be explained with the help of fossils of fish, frog and fresh water tortoise. Such beds are exposed at Malabar hill (Mumbai) and indicate that these fossiliferous beds were formed under fresh water conditions (probably a lake) and are called as intertrappean beds.

(e) Fossils have also helped geologists in identifying and distinguishing between rocks of different origin having similar appearance. For example, the beds that are present below the Deccan Traps near Jabalpur (Madhya Pradesh) are either sedimentary (called as Lameta beds) or metamorphic rocks (Archaean gneisses and schists). Both have been calcified by the percolating ground water solutions from the overlying Deccan Traps. Even though, schists, gneisses and Lametas look alike only Lametas possess fossils and can be easily distinguished.

(f) The modern day discoveries of coal, petroleum and mineral ores have been possible with the help of fossil studies. The microfossils like foraminifera have been useful in identifying petroliferous horizons in Texas (USA). Likewise, the coal seams have always been store houses for a large variety of plant fossils. Some of the mineral and ore deposits have been associated with the presence of typical species of organisms.

2.   In determining the depositional environments: As stated earlier, the presence of animal and plant fossils in sedimentary rocks always indicates the type of environment that existed during the sedimentation. For example, the presence of corals will always indicate tropical or sub-tropical climate at the time of their formation. The presence of vertebrates and flowering plants is suggestive of terrestrial conditions. The fossil of dinosaur *Mesosauras* suggests fresh water environments while the fossil of *Eohippus* (most primitive fossil of horse) is an indication of terrestrial environments. The fossils belonging to the phylum Cephalopoda (e.g., nautilus and ammonoid) indicate marine environment. The admixture of the marine and terrestrial fossils, however, will be the characteristic of shallow water marine conditions. The nature of Lameta and intertrappean beds can be distinguished easily with the help of their fossil contents. The marine environment of Lametas is evident from the presence of fossils of nautilus, gastropods and lamellibranchs while estuarine environment of fossils in inter-trappean beds is proved by the presence of fossils of frog and fresh water tortoise and total absence of marine fossils like corals, cephalopods and echinoderms.

3.   In Stratigraphy: The study of the layered (stratified) rocks of the earth constitutes the science of stratigraphy. It deals with the formation of the stratified rocks, their age and time interval taken by each of their individual member for its formation. In addition, these rock formations are also correlated with their nearby, as well as, distant formations with similar affinities all over the world. This study is called as stratigraphic correlation. In such correlations, the fossils play an important role of being a geologist's clock because they are able to provide vital information about the nature of the environments that existed in the geological past. The usage of fossil record in stratigraphy can be explained as given below:

(a)   A particular rock sequence may contain characteristic group of fossils so that the rocks themselves can be described as lamellibranch limestone, gastropod limestone or coralline limestone.

(b)   The study of the worldwide correlation of vertical stratigraphic column in terms of time and space constitutes the branch of geology called chronostratigraphy. The geological events are best understood when they are explained on the basis of fossil contents. The oldest Pre-Cambrian Eon has been subdivided into three Eons on the basis of their fossils contents. The Hadean Eon does not have any diagnostic record; the Archean Eon contains fossils of only single-celled organisms while the Proterozoic Eon bears multicellular and more evolved organisms.

(c)   The fossils also provide information about the age of the rocks. Accordingly, the rocks that are geologically older possess primitive and simpler fossils while the younger rocks have more complex and advanced fossil assemblage. Therefore, based on the type of fossils the rock possesses, the relative age of the rock can be estimated. For instance, the age of the Deccan Trap has been decided on the basis of fossils that the infratrappean beds contain. The algal remains from these beds have not been observed in beds older than Paleocene epoch (66 million years) in any part of the world. Hence, as infratrappean beds are present below the Deccan Trap flows, it is inferred that the age of Deccan Traps cannot be older than Palaeocene.

(d)   Some fossils are so typical of certain stratigraphic horizons that they are used as index or guide fossils of that particular geological age. These fossils are widely distributed in geographical regions but are restricted to only specific geological time range. For example, the plant fossil of *Glossopteris* and *Gangamopteris* are index fossils of Permain age. The fossil of Calceola (Phylum *Coelenterata*) is typical of Devonian period and so on.

(e)   Based on the index fossils and other textural characteristics, the stratigraphic successions can be divided into different zones. Such a type of classification is called as biostratigraphic classification. The important fossils used in this type of classification are called as zone fossils. Some of the important zone fossils are *Ophioceras sakuntala* in Triassic rocks of Spiti valley (Himachal Pradesh) and *Tropites subbullatus* in Kashmir.

4.   In Palaeontology: In Palaeontology, the fossil helps in unfolding the mystery of evolution of life on the earth and the diversification of different living species. In this regard, the present day organisms are considered as the descendants of those which are recovered as fossils. Not all the fossils resemble their living and more evolved counterparts but the study of their evolution is simplified with the help of fossils as given below:

(a)   The study of nautilus in the modern oceans has helped to understand the anatomy of fossil ammonoid which is not living at present (extinct fossil).

(b)   The nature of continuous and gradual evolution of animals as stated by **Darwin** (The Theory of Organic Evolution) can be explained with the help of fossils. Such studies have shown that the horse that we ride today had its ancestors in the form of a miniature sized mammal. Even though, it has only one hoof today, the oldest ancestor of the horse was a multi-toed, dog-sized animal in Eocene period (37-58 m. years ago). Fossils of the horse as observed in the sediments of different ages and their relative sizes indicative of their gradual evolution is illustrated in Fig. 2.4.

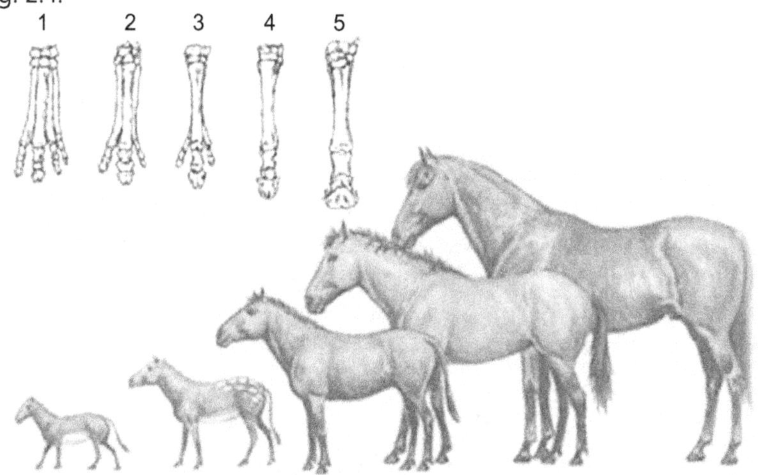

1. Eohippus   2. Mesohippus 3. Merychippus 4. Pliohippus   5. Equus

37 to 58 million years ago   **Present Day horse**

Fig. 2.4: Evolution of different sized fossils of horse

(c)   It is because of the study of different fossils that we are able to establish a link between living species. Based on the record of the changes that organisms have undergone, it was possible to establish the family tree or evolutionary pattern of living organisms. For example, the structure of dinosaurs was visualised only after comparing the fossil bones with the modern day reptiles.

(d)   The studies of fossil skin impression of woolly mammoth were useful in re-creating the other soft parts of the extinct animal.

5. As Climatic Indicators: Different types of fossils have been characteristic of certain climatic conditions. Their study has not only made the reconstruction of the earth's surface in the past possible but has also unfolded the mystery of unusual rocks in the modern day climates. For example, the coral reefs have been identified in the New Siberian Islands in the Arctic Circle and coal beds, with typical plant fossils of tropical climates, have been recovered at Antarctica. These are the signs of the existence of totally different climatic conditions of the past.

6. In Research: The fossils have helped scientists to visualise or reconstruct the geophysical and environmental conditions of the past. The study of dinosaurs and the extinct flying reptiles has enabled scientists to predict the geophysical parameters of the then existing earth's surface. These studies have revealed that the gravity of the earth during Cretaceous was lower than present. The bulkiness of the dinosaurs and the enormous wing span of the flying reptiles are suggestive of weaker gravity during Cretaceous period.

The corals produce growth bands, which are synchronized with the solar days of the modern day solar system. Hence, the modern corals possess bands amounting to approximately 360 which match with the present solar year of the earth. But, the study of Ordovician corals reveals that they have 412 bands while Silurian corals have 400. This explains that the earth did not have its rotation around the sun in the same manner as it has today.

POINTS TO REMEMBER

- Palaeontology involves the study of dead remains of microscopic and megascopoic animals and plants that once lived on the earth's surface and are now a part of the stratified rocks.
- Chevalier de Lamarck has been honoured as the *"Father of Invertebrate Palaeontology"* .
- The branches of Palaeontology are palaeobiology, micropalaeontology and ichnology
- Palaeobiology is the branch of palaeontology that deals with the study of plant and animal remains in the geological sediments of the past.
- Micropalaeontology is the study of microfossils.
- Ichnology is a branch of palaeontology that deals with the study of tracks, trails and burrowing habits of organisms.
- The main characteristic of any fossil is its preservation in the sediments.
- The process of fossilization is referred to as taphonomy.

- Fossils have helped geologists in identifying and distinguishing between rocks of different origin having similar appearance.
- Fossils play an important role in determining the depositional environments.
- The study of the stratified rocks of the earth constitutes the science of stratigraphy.
- It deals with the formation of the stratified rocks, their age and time interval taken by each of their individual member for its formation.
- In Palaeontology, the fossil helps in unfolding the mystery of evolution of life on the earth and the diversification of different living species.
- Different types of fossils have been characteristic of certain climatic conditions.
- The fossils have helped scientists to visualise or reconstruct the geophysical and environmental conditions of the past.
- In spot sampling, the sediments to be collected are of predetermined stratigraphic levels.
- In channel sampling is undertaken for large stratigraphic sequences with variation of lithology and vertical intervals ranging from 1 to 3 mts.

EXERCISE

1. Define the term fossil and describe the conditions necessary for fossilisation.
2. Define the term palaeontology and describe the branches of paleontology.
3. Which rocks are likely to possess fossils? Describe how different sites are selected for fossil collection.
4. Describe various uses of fossils in geology.
5. Write notes on:
    (a) Modes of preservation of fossils
    (b) Mould and cast
    (c) Petrification
    (d) Carbonisation
    (e) Evolution of fossils
    (f) Body and trace fossils
    (g) Leaching and Permineralisation
    (h) Fossils and past environment

Chapter 3...

PHYLUM ECHINODERMATA

Contents ...

SYSTEMATIC POSITION

Kingdom	Animalia
Sub-Kingdom	Invertebrata
Phylum	Echinodermata (*Echino* – spiny, *dermata* – skin)
Sub Phylum	(a) Elutheroza (Free moving forms)
	(b) Pelmatozoa (attached forms)
Class	Echinoidea
Subclass	(a) Regularia
	(b) Irregularia

3.1 INTRODUCTION

The phylum Echinodermata includes exclusively marine animals which are highly organised. These are present abundantly in the rocky regions of the sea floor and rather sparsely on the sandy sea floor. These are excellent scavengers and feed on the remains of dead organisms.

The name Echinodermata is derived from the Greek word *echinos* - spiny, *dermata* - skinned. The Echinoderms in the present day modern oceans are the starfish, sea-urchins, sea-cucumbers and sea lilies. They are famous as food delicacies all over the world.

The echinoderms are divided into two main groups viz

(a) Those which are not attached but are free moving (Eleutherozoa) and (b)　Those which are more or less permanently fixed in the mode of growth (Pelmatozoa).

Most echinoderms are adapted to the saline waters of the shallow as well as deep oceans. They rarely live in brackish water and are not found in fresh water.

The echinoids are described as (1) regular and (2) irregular, depending upon their shape and symmetry. The regular echinoids are globular in shape and exhibit a radial pentamerous symmetry whereas the irregular echinoids are generally heart-shaped and possess a bilateral symmetry.

3.2 MORPHOLOGY OF THE HARD PARTS

Echinoderms usually possess an internal mesodermal skeleton which is made of calcite. In living conditions, this mesodermal skeleton is covered by a thin protoplasmic skin. As the skeleton is of calcite, it is easily preserved in carbonate sediments. Echinoids are globular, discoidal or heart shaped. The globular test of a typical echinoid is about 10 centimeters in diameter and is slightly flattened at the poles (Fig. 3.1). In living conditions, the skeleton is covered with spines, which are 1 to 2 cm in length.

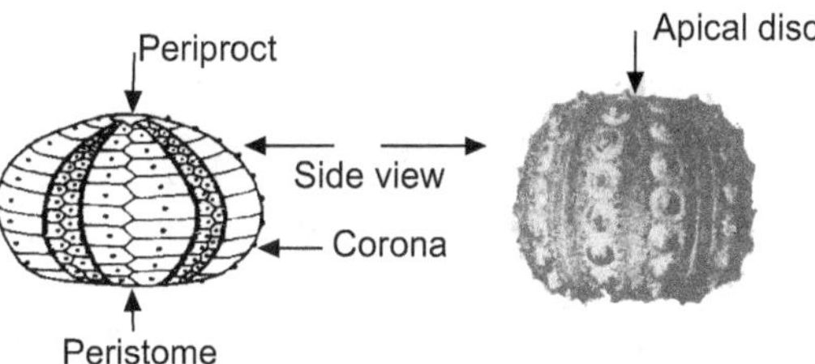

Fig. 3.1: Morphology of a typical Echinus

The dorsal part (upper pole) of the test in which the anus is situated is called as the aboral side of the animal. The ventral side (lower pole) where the mouth of the animal is located forms the oral side. The test of a typical echinoid consists of three major parts viz. the apical disc, the corona and the peristome (Fig. 3.1).

The Apical disc: On the aboral surface, there is a flower shaped structure called as the apical disc. It consists of a double ring of plates surrounding a central periproct, where the anus of the animal is situated (Fig. 3.2).

The larger plates surrounding the periproct are called the genital plates. Each genital plate is perforated by a genital pore, through which the eggs or sperms are discharged. The genital pores of the female echinoids are larger in size than the pores of the male echinoids. Hence, these characteristics help to determine the sex of the fossil echinoids. One genital plate is larger than the other genital plates and bears numerous microscopic pores. It can easily be identified and is termed as the madreporic plate. This plate regulates the water – vascular system of the animal. The smaller ocular plates alternate with the genital plates. Each ocular plate bears a small ocular pore which also regulates the water-vascular system.

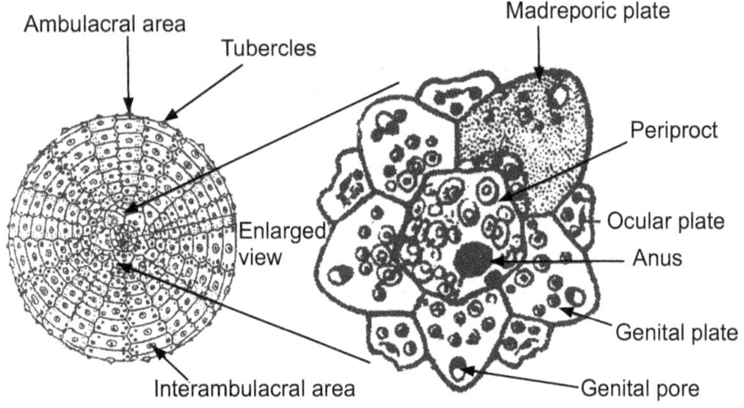

Fig. 3.2: The apical disc of Echinus

3.3 VARIATION IN THE APICAL DISC

The apical discs in most regular echinoids are large, e.g., *Cidaris*, *Peltastes* and *Salenia*. In few regular echinoids, the genital plates are completely separated by smaller oculars so that a row of ten plates surrounds the periproct. Whenever the oculars separate the genitals and touch the periproct, they are called as *insert* and when they do not touch the periproct, they are called *exert* (Fig. 3.3a and b) respectively.

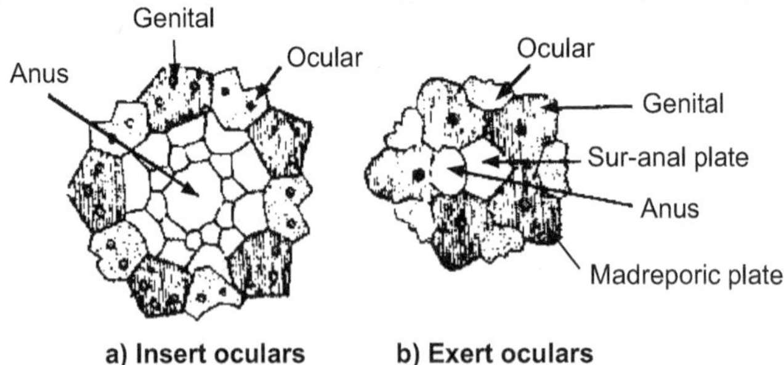

a) Insert oculars   **b) Exert oculars**

Fig. 3.3: Variation in the apical discs

In most cases, each genital plate has usually one perforation but in many echinoids of the Palaeozoic forms, there are two, three or more perforations. In some species, e.g., *Salenia* and *Peltastes*, there is an extra plate in the apical disc, commonly known as, the sur-anal plate (Fig. 3.3b).

In the case of irregular echinoids, the apical disc is small as it does not enclose the periproct. The madreporic plate, in some cases, may extend to the centre of the disc (e.g., *Conulus*, Fig. 3.4a) or may reach the posterior borders, separating the posterior oculars (e.g., *Spatangus*, Fig. 3.3b). In some species, e.g., *Collyrites*, the apical disc is considerably elongated by virtue of which the two posterior oculars get separated from the rest of the apical disc by a chain of smaller plates (Fig. 3.4c). In some cases, the genital plates are found to be fused together, e.g., *Clypeaster* (Fig. 3.4d). In *Echinocorys* and *Holaster*, the apical disc is found to be elongated whereby the anterior genitals are separated from the other genitals by two oculars, which join in the middle and the posterior genitals are absent (Fig. 3.4e).

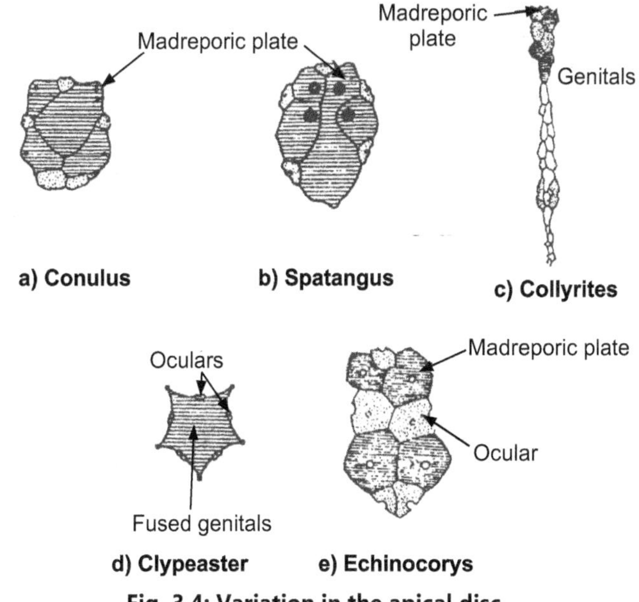

Fig. 3.4: Variation in the apical disc

Corona

The region between the apical disc and the peristome forms the corona of the test. It consists of ten double rows of plates; the larger ones which arise from the genital plates and end up at the peristome are called as the **interambulacral** plates. Each interambulacral plate bears a shallow, smooth pit, the areole, within which is a conical boss surrounded by a globular mamelon (Fig. 3.6a).

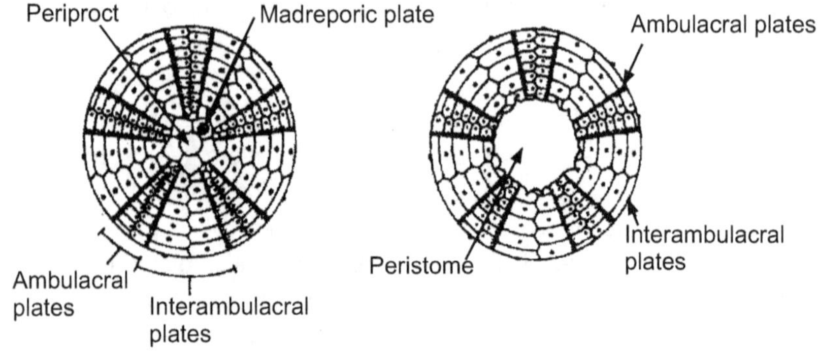

a) Dorsal view   **b) Ventral view**

Fig. 3.5: The corona of a typical echinoid test

The boss and its mamelon form a primary tubercle carrying primary spines of around 10 cm in length, which serve as defense mechanisms of the animal (Fig. 3.6 b).

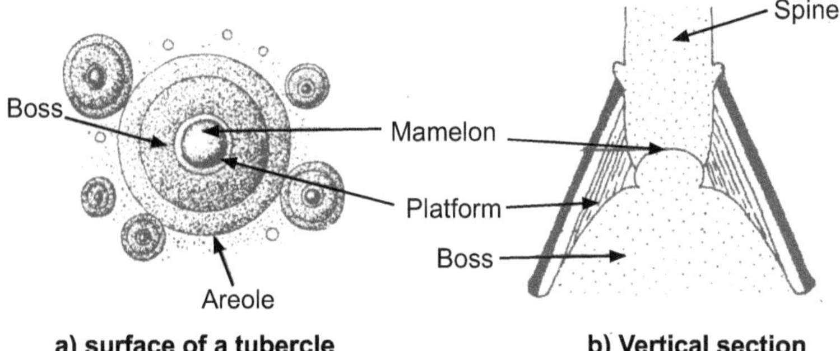

a) surface of a tubercle   **b) Vertical section**

Fig. 3.6: Tubercles and spines in the echinoids

When the animal dies, the muscles holding the spines decay, thereby, liberating the spines from the main skeleton. Thus, the spines are not preserved with the main skeleton. However, the smaller tubercles (lacking the areole) are found scattered all over the surface of the interambulacral plates and are known as secondary tubercles.

The **ambulacra** are five narrow segments which originate from the ocular plates and end up at the peristome. Each ambulacral plate bears three sets of paired pores near the outer edge of the plate. These pore-pairs are the sites through which the tube feet emerge from the internal part of the water – vascular system. Both ambulacra and interambulacra consist of double columns of elongated calcitic plates which meet towards the center in a zigzag pattern.

The **Peristome:** The oral (bottom) side of the animal is termed peristome. The peristome is a large area covered when living by a flexible plated membrane which surrounds the central circular mouth (Fig. 3.5b). Five pairs of gill notches are found where the interambulacra meet an edge of the peristome. The peristome being delicate is rarely preserved with the fossil.

3.4 GEOGRAPHICAL AND GEOLOGICAL DISTRIBUTION

The first known occurrence of echinoderms are found in the lower Cambrian rocks. However, they started diverging into widely varied forms from the Ordovician period. The Palaeozoic rocks show the dominance of the attached echinoderms (Palmatozoa). The free moving echinoids (Eleutherozoa) were found to be in much abundance in the post-Palaeozoic deposits. However, at the end of the Permian period, all the echinoderms suffered a drastic reduction in numbers. The first irregular echinoid appeared during the Jurassic period. The abundant echinoids found in the modern seas had their flourishing existence form the end of the Triassic period.

POINTS TO REMEMBER

- The phylum Echinodermata includes exclusively marine animals which are highly organised.
- They are excellent scavengers and feed on the remains of dead organisms.
- The echinoderms are divided into two main groups viz ones that are free moving (Eleutherozoa) and ones that are more or less permanently fixed in the mode of growth (Pelmatozoa).
- The echinoids are described as regular and irregular, depending upon their shape and symmetry.
- The regular echinoids are globular in shape and exhibit a radial pentamerous symmetry.
- The irregular echinoids are generally heart-shaped and possess a bilateral symmetry.
- Echinoderms usually possess an internal mesodermal skeleton which is made of calcite.

- Echinoids are globular, discoidal or heart shaped.
- The test of a typical echinoid consists of three major parts viz. the apical disc, the corona and the peristome.
- On the aboral surface, there is a flower shaped structure called as the apical disc.
- Each genital plate is perforated by a genital pore, through which the eggs or sperms are discharged.
- The region between the apical disc and the peristome forms the corona of the test.
- The ambulacra are five narrow segments which originate from the ocular plates and end up at the peristome.
- The oral (bottom) side of the animal is termed peristome.
- The peristome is a large area covered when living by a flexible plated membrane which surrounds the central circular mouth.

EXERCISE

1. Describe the morphology of the hard parts of a typical echinoderm.
2. Differentiate between the regular and irregular echinoids. Give their geographical and geological distribution.
3. Write notes on
   (a) Variation in the apical disc
   (b) Corona and peristome.

Chapter **4**...

PHYLUM ARTHROPODA

Contents ...

SYSTEMATIC POSITION

Kingdom	Animalia
Sub-Kingdom	Invertebrata
Phylum	Arthropoda (animals with jointed feet)
Class	Crustacea (animals with jointed external skeleton)
Order	Trilobita (animals with three lobed exoskeleton)

4.1 INTRODUCTION

The existence of the animals belonging to the phylum Arthropoda has been recorded right from the Cambrian period to the present. The name Arthropoda in Greek, is translated into *arthro* - jointed, *podos* - foot, as these animals possess an elongated and segmented body with paired, jointed limbs or appendages. Arthropods are found in both marine and fresh water environments; in the terrestrial and aerial habitat. Arthropods are bilaterally symmetrical and their body consists of a number of articulating segments, each bearing a pair of limbs which have different functions. The exoskeleton is flexible over the joints between the segments facilitating free movement. They have a chitinous external skeleton.

The Arthropods have a highly developed nervous system and a brain from which the ventral nerve cords arise and give off branches to each segment. The heart brings about the blood circulation. The eyes, if present, are either simple or compound. Respiration takes place through the gills or through the surface of the body. The Arthropods belonging to the order Trilobita were found only in the marine environment and are totally extinct today.

4.2 MORPHOLOGY OF THE HARD PARTS

Trilobites have been assigned the status of a phylum because of the segmented body and paired jointed limbs. The exoskeletons of these animals were probably chitinous impregnated with calcium salts. This formed the supportive shield which provided a firm surface of attachment for the muscles and also prevented the desiccation of the animal. Just like the cockroaches, the trilobites also moulted from time to time and hence, many exoskeletons were given off by the animal in its life time. Even though, the animals are extinct, their exoskeletons have been preserved as fossils. In most cases, the fossils are the dorsal exoskeletons but preservation of the ventral side of the body has also been recorded.

The protective dorsal exoskeleton has a typical flat or gently convex shape with its margins turned inward so as to partially cover the ventral side to form a rim. On the basis of the dimension, it appears that these animals measured between 0.5 to 70 cms in length.

The body of the animal can be divided into three main parts viz. cephalon (head), thorax (body) and pygidium (tail) as shown in Fig. 4.1. The exoskeleton covering the dorsal side of the animal is called the carapace. The exoskeleton is grooved longitudinally by two axial furrows, which separate an arched central region. This axis forms two side regions, the pleurae.

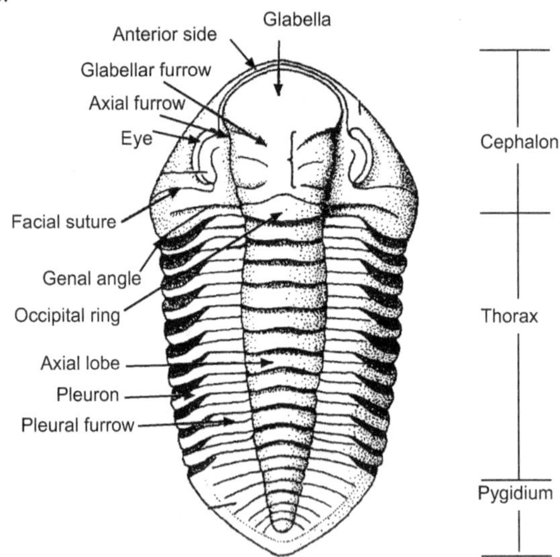

Fig. 4.1: Morphology of a trilobita on its dorsal side

The Cephalon (Head): The anterior parts consist of six fused segments forming the head shield or cephalon. The cephalon is typically semicircular in outline. Its anterior side is semicircular in outline while the posterior end is almost flat. The cephalon is divided into two major portions viz. the central raised portion called as the glabella; and two lateral extensions called the cheeks. The two cheeks are separated by well defined axial furrows. The glabella is generally convex and exhibits an incomplete segmentation in the form of glabellar furrows. However, the posterior most furrow is well defined, axial furrow is well defined and as it encircles the cephalon it is often called as the occipital ring.

The cheeks are present on either side of the glabella as triangular extensions that may be continuous in front of the glabella or may be separated. Each cheek is crossed by a suture known as the facial suture. The facial suture defines two regions viz, the fixed attached to it; and the free cheeks, which are slightly movable and possess compound eyes. The eyes are elliptical or conical in shape and their presence on the free cheeks is indicative of the viewing ability of the animal. The eyes were the first ever sensory organs of the living organisms.

The posterior extensions of the cheeks on the lateral sides form the genal angles. These genal angles may be either rounded as in the case of calymene or pointed and produced into spines as in the case of paradoxides.

The Thorax: The mid-portion of the animal consists of a number of freely articulated segments and is called as the thorax. It usually consists from two to forty-two segments which resemble each other except that they become gradually narrower towards the pygidium. Each segment is divided longitudinally into three portions by two lateral sides are called as pleurae. The pleurae may have rounded or pointed ends. Due to the flexibility of these thoraxic segments, the animal could roll in such a way that its head and tail shields would get pressed together to safeguard its soft, ventral region.

The Pygidium : The tail or the posterior most part of the animal is called as the pygidium. It is semicircular or triangular and is made up of 2 to 30 fused segments. As in the case of the thorax, the pygidium is also divided into three regions, viz, the central axial and the lateral pleural regions. It is believed that the pygidium served as a propeller for the movement of the animal through the water.

Limbs of the Trilobites: Except the posterior most segments, all the other segments possessed a pair of appendages or limbs. These appendages were used by the animal for swimming or crawling along the sea floor. Fossils of these limbs have been preserved only in suitable conditions in the fine grained sediments.

4.3 GEOGRAPHICAL AND GEOLOGICAL DISTRIBUTION

These exclusively marine organisms evolved during the Cambrian and the Ordovician periods. However, during the Silurian period, their population started declining. This decline has accentuated during the Devonian period due to the incoming fishes. Ultimately, the Permo-Carboniferous time was the culminating period for the Trilobites when superior fishes dominated the marine waters.

POINTS TO REMEMBER

- Arthropods possess an elongated and segmented body with paired, jointed limbs or appendages.
- The Arthropods have a highly developed nervous system and a brain from which the ventral nerve cords arise and give off branches to each segment.
- They are found in both marine and fresh water environments; in the terrestrial and aerial habitat.
- The exoskeletons of Trilobites were probably chitinous impregnated with calcium salts.
- The body of the animal can be divided into three main parts viz. cephalon (head), thorax (body) and pygidium (tail).
- The exoskeleton covering the dorsal side of the animal is called the carapace.
- The anterior parts consist of six fused segments forming the head shield or cephalon.
- The mid-portion of the animal consists of a number of freely articulated segments and is called as the thorax.
- It usually consists from two to forty-two segments which resemble each other.
- The tail or the posterior most part of the animal is called as the pygidium.
- It is semicircular or triangular and is made up of 2 to 30 fused segments.

EXERCISE

1. Describe briefly, the morphology of the hard parts of the Trilobites, with the help of neat labelled diagrams.
2. Write notes on:
   (a) The Cephalon   (b) The Thorax
   (c) The pygidium
   (d) Geographical and geological distribution of Trilobites

Chapter **5**...

PHYLUM MOLLUSCA (CLASS LAMELLIBRANCHIA)

Contents ...

SYSTEMATIC POSITION
CLASS LAMELLIBRANCHIA (PELECYPODA)

Kingdom	Animalia
Sub-Kingdom	Invertebrata (without back bone)
Phylum	Mollusca (soft-bodied animals)
Class	Lamellibranchia or Palecypoda
Sub-Class : 1.	Palaeotaxodanta
	(Regular hinge teeth arrangement)
2.	Isofilibranchia
	(Gills occur as sheets of filaments)
3.	Heteroconchia
	(Variable and complex hinge teeth arrangement)
4.	Anomalodesmata
	(Valves joined with internal ligament and absence of hinge teeth)

5.1 INTRODUCTION

The members of the class lamellibranchia or pelecypoda are exclusively aquatic animals that are found in fresh, marine and brackish environments. They are, however, more abundant in marine environment. Some are slow moving bottom dwellers (benthos), others are attached, while a few lead their life as swimmers (nektonic). The two basic modes of life are (i) infaunal i.e., burrow-dwelling and (ii) epifaunal i.e., surface-dwelling. Those which remain attached are characterised by irregular shapes as compared to those which burrow themselves in the sediments. Few lamellibranchs such as *Mytilus* have thread-like features called byssus threads, by which they anchor themselves to the rocks. Some authors propose the term lamellibranch (meaning animals with a pair of leaf-like gills) for this class, while others prefer the term pelecypoda (meaning those possessing a hatchet-shaped foot). These aquatic animals feed on the suspended food particles from the water column. The food intake is facilitated by the presence of an inhalant siphon and the spent water is drawn out through the exhalant siphon.

5.2 MORPHOLOGY OF THE HARD PARTS

Pelecypoda shells consist of two convex and hard calcareous valves (hence, the term bivalved shells), which are placed on the right and left sides of the animal body. These shells consist of three layers, viz. the outermost periostracum made up of a chitinous layer, a middle layer of ostracum made up of prismatic calcium carbonate and the innermost layer of hypoostracum, which is a pearly layer. Each valve is drawn into a beak like portion called the umbo which usually point towards the anterior part of the shell (Fig. 5.1).

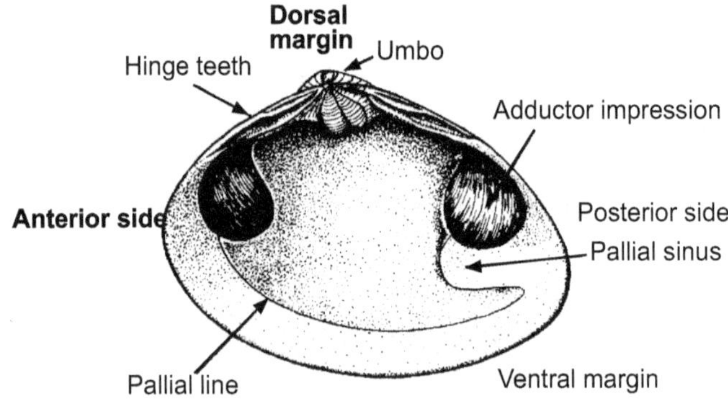

Fig. 5.1: Morphology of a bivalve shell

The pelecypods, in which the umbo points towards the anterior end, the shells are termed prosogyrate e.g., *Meretrix*, whereas in those where it points towards the posterior end of the shell, are termed as ophisthogyrate e.g., *Nucula*. In *Pecten*, the umbo forms a wing-like projection (Fig. 5.2) and the shape of the valve is arcuate; while the shapes of other shells include triangular (e.g., *Trigonia*) and oblong (e.g., *Mytilus*).

5.3 SYMMETRY OF THE SHELL

Even though, most of the lamellibranchs are inequivalved (e.g., oysters and scallop), the size of the shell ranges from a fraction of a centimeter to gigantic shells with a width of sixty centimeters, a length of about two meters and weighing several hundred kilograms.

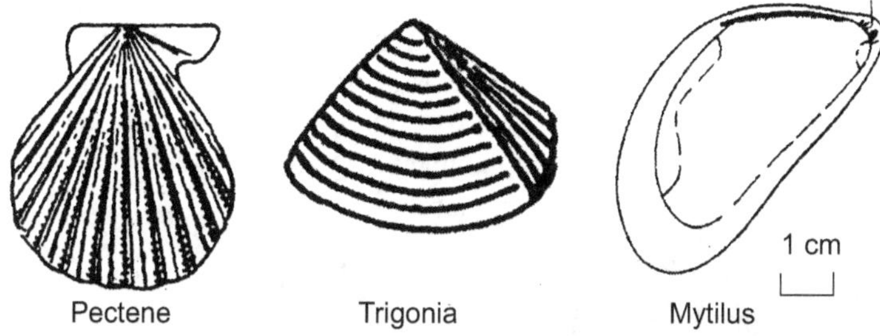

Fig. 5.2: Shapes of bivalves

Both the valves are articulated and held together by an elastic ligament and hinge situated on the dorsal side of the shell. The lower margin along which the shell opens is called the ventral margin. If a line is drawn from the umbo to the ventral margin, in most cases, it does not divide the shell into two equal parts and therefore, the shell is inequilateral as shown in Fig. 5.3.

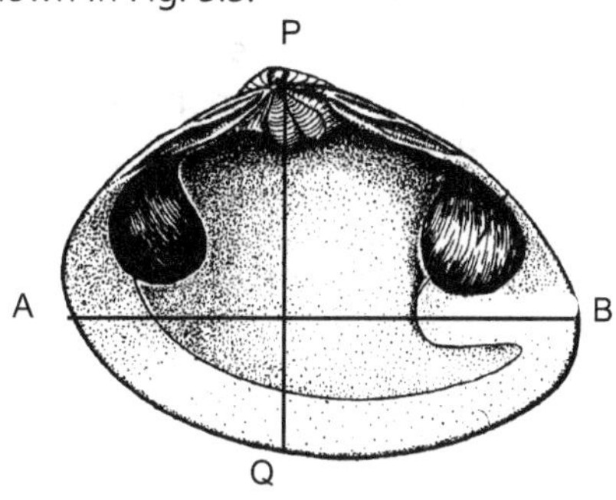

Fig. 5.3: Symmetry of a bivalve shell

The distance between the anterior and the posterior extremities (A-B) is the length of the valve, while the distance P-Q (between the dorsal and the ventral margin) gives the height of the shell. The distance between the two normals drawn parallel to the valves (M–N) measures the thickness of the shell (Fig. 5.4a).

When the two valves are intact and viewed dorsally, in front of the umbons, an oval shaped depression, known as lunule, is seen (Fig. 5.4 b). There is a similar larger area of depression behind the umbons, called the escutcheon.

When the animal dies, all the soft parts including those which were attached to the inner portion of the shell, leave behind their scars. In living pelecypods, there are usually two muscles which are used for closing the shell. These are called as adductor muscles. The impressions of these muscles are termed as anterior adductor impression, when located on the anterior side and as posterior adductor impression, when located towards the posterior side.

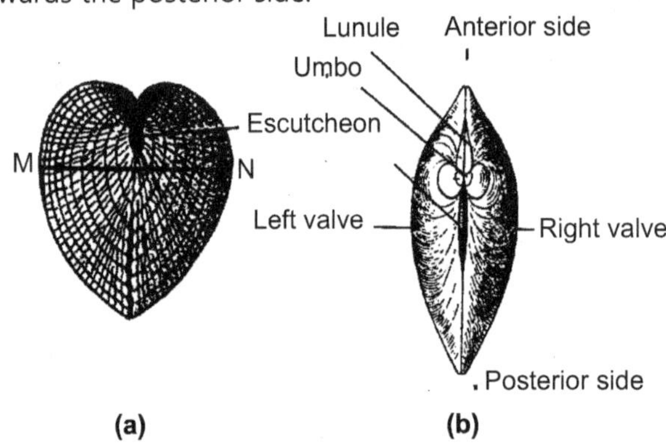

Fig. 5.4: Thickness of a pelecypod shell

When there are two adductor impressions, the shell is called as a dimyarian shell. In some cases, the two adductor impressions are nearly equal in size and similar in shape. Such a shell is termed as homomyarian (e.g., *Meretrix*). In some Pelecypods, the anterior adductor impression is smaller in size than the posterior impression and such shells are said to be heteromyrian (e.g., *Arca*). Pelecypods in which only a single adductor impression is observed, the shell is termed as monomyarian (e.g., *Pecten*). Along the ventral margin of the shell, a scar or line is seen extending from the anterior muscle impression to the posterior muscle impression. This line is called as the mantle or pallial line. The pallial line, in most cases, is simple or entire e.g., *Arca*, though many pelecypods have a pallial line with a deep indentation, commonly referred to as a pallial sinus e.g., *Meretrix*.

5.4 ORNAMENTATION

The surface of the bivalve shell may be smooth, or may be ornamented with rediating or concentric ribs (Fig.5.5) and striae, or with tubercles, or spines (Fig. 5.6).

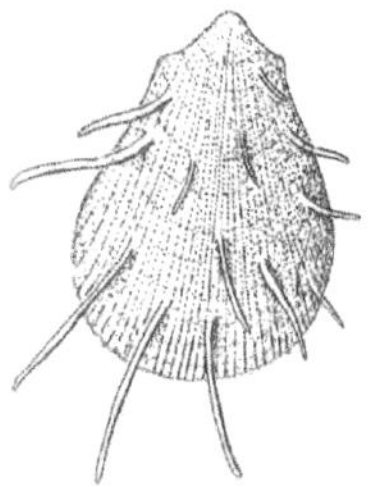

Fig. 5.6: Spines on lamellibranch shell

Often are seen the concentric lamellae, which represent periods of growth. The part of the shell at the umbo is that which was first formed, and often differs in ornamentation and form from the other parts. The margins for the valves may be smooth or crenulated. Sometimes, as in some species of pecten, the entire shell may be corrugated. In many genera, the two valves can be completely closed, in others they are always open at some part, and are then said to be gaping. This gap occurs most frequently at the posterior end and serves for the passage of the siphons. An anterior gap may also be present.

5.4 HINGE PLATE

The dorsal end of the valve displays the hinge plate. The hinge plate bears teeth and between the teeth are the sockets. The teeth occur as elevations, while the sockets are in the form of depressions. The teeth and sockets act as guides ensuring that the two valves close perfectly giving a tight fit. In the pelecypods, the teeth and sockets occur in both the valves. The dorsal margin of the valve on which the teeth and sockets occur is known as the hinge line. Generally, it is curved as in *Meretrix* or it is straight, as in *Arca* (Fig. 5.7).

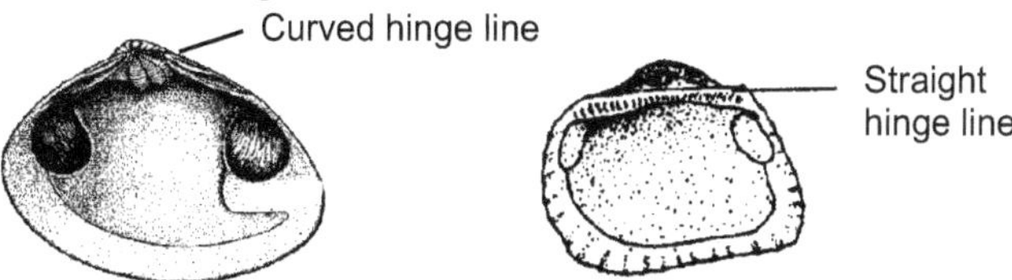

Fig. 5.7: Hinge line in Meretrix and Arca

In living pelecypods, there is a rubber – like substance known as the ligament which not only connects the two valves but is also used for opening the shell. It lies in an elongated pit, posterior to the umbo but is seldom preserved.

Pelecypods exhibit several kinds of dentation and hinge structures. Of these, only a few are clearly recognised. Some of the important types of dentation are illustrated in Fig. 5.8 (a-g) and described on the next page.

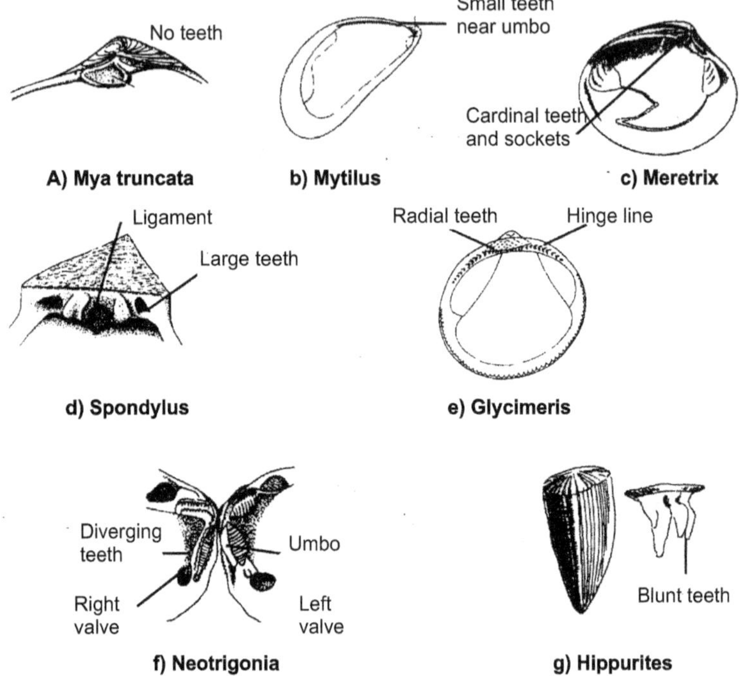

Fig. 5.8: Types of dentation in pelecypods

(a) **Desmodont:** In such shells, the teeth are much reduced or are absent along the hinge structure. However, accessory ridges are present along the hinge margin e.g., *Myatruncata.*

(b) **Dysodont:** These shells have small and simple teeth which are present near the edge of the valve e.g., *Mytilus.*

(c) **Heterodont:** In these shells, two different types of teeth are present. Two or three massive cardinals are present below the umbo and a narrow, elongate and lateral tooth is placed away from the umbo e.g., *Meretrix.*

(d) Isodont: These shells are characterised by very large teeth which are located on either side of a central ligament pit e.g., *Spondylus*.

(e) Schizodont: The teeth are few in number and they form grooves which diverge below the umbo e.g., *Neotrigonia*.

(f) Taxodont: The teeth are numerous and sub-parallel or are radially arranged along the entire hinge line e.g., *Glycimeris*.

(g) Pachydont: Due to the attached mode of life of this animal, its teeth are very large, heavy and blunt e.g., *Hippurites*.

POINTS TO REMEMBER

- The members of the class lamellibranchia or pelecypoda are exclusively aquatic animals that are found in fresh, marine and brackish environments.

- The two basic modes of life are (i) infaunal i.e., burrow-dwelling and (ii) epifaunal i.e., surface-dwelling.

- Pelecypoda shells consist of two convex and hard calcareous valves (hence, the term bivalved shells), which are placed on the right and left sides of the animal body.

- These shells consist of three layers, viz. the outermost periostracum made up of a chitinous layer, a middle layer of ostracum made up of prismatic calcium carbonate and the innermost layer of hypoostracum, which is a pearly layer.

- Each valve is drawn into a beak like portion called the umbo.

- The pelecypods, in which the umbo points towards the anterior end, the shells are termed prosogyrate.

- If a line is drawn from the umbo to the ventral margin, in most cases, it does not divide the shell into two equal parts and therefore, the shell is inequilateral.

- The dorsal end of the valve displays the hinge plate.

- Some of the important types of dentation are desmodont, dysodont heterodont, sodont, and schizodont, taxodont, pachydont.

- All bivalves are aquatic and occur today in marine, fresh water, as well as, in some brackish-water environments.

EXERCISE

1. Describe the morphology of hard parts of the pelecypods with special reference to their external and internal morphology.

2. Differentiate between the pelecypods and brachiopod shells.

3. Describe the length, height, thickness and forms of pelecypod shells.

5. Write notes on

    (a) Adductor muscles in pelecypods

    (b) Symmetry of a pelecypod shell

    (c) Ornamentation

    (d) Hinge plate

    (e) Types of hinge lines

Chapter **6**...

CLASS GASTROPODA

Contents ...

SYSTEMATIC POSITION

Kingdom   Animalia

Sub-kingdom Invertebrata

Phylum   Mollusca

Class   Gastropoda

Subclass   1. Protogastropoda

(Mostly marine, includes planispirally coiled shells that are now extinct, e.g., *Scenella*)

2. Prosobranchia

(Marine shells with simple gill structure. Mantle cavity is fully twisted, e.g., *Turritella, Conus*)

3. Opisthobranchia

(Marine shells with complex gill structure. Shells have undergone distortion and appear as untwisted loop, e.g., *Limacina*)

4. Pulmonata

(Terrestrial or fresh water gastropods with or without shell, respiration with the help of lungs, e.g., *Planorbis*)

6.1 INTRODUCTION

Gastropoda (in Greek *Gastros* - stomach, *Podos* - foot) is a major class of the Phylum Mollusca. It includes animals which bear a coiled (e.g., *Turbo*) or uncoiled (e.g., *Patella*) calcareous shell, while others like slugs have no hard parts. The Gastropods are the most diversified group amongst the molluscs and occur in both aquatic (fresh and marine water) as well as terrestrial environments. They are mostly marine and are confined to the shallower parts of the sea, e.g., benthonic habitat. However, some have taken refuge in the oceanic depths more than five kilometers deep, while some swim in the near-surface waters, i.e., they have a nectonic habitat. Geologically, they are found in sedimentary rocks of all ages and their presence as living animals, as well as, fossils helps the palaeontologists to unravel the geological environments which existed in the past.

6.2 MORPHOLOGY OF THE HARD PARTS

The gastropod shells are univalved as they consist of only one valve. Usually the shells consist of a long tube which is broad and open at one end and gradually tapers towards the other end. However, in the case of *Patella*, it occurs as a hallow cone. The average diameter of a gastropod shell is 2.5 cms and its length varies between 2 to 15 cms. However, microscopic gastropods (pteropods) have also been recorded. Compositionally, the shells are aragonitic calcareous.

The shell is coiled spirally into a screw like structure and each of the individual coils is described as a whorl (Fig. 6.1).

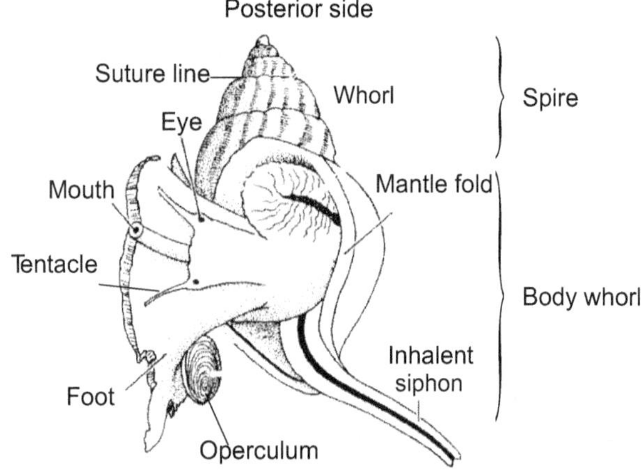

Fig. 6.1: Morphology of Gastropod shell

The shell is composed of many whorls. All the whorls are in contact with each other and the dividing line between two whorls is the suture line. *Siliquaria* and *Vermetus* have whorls which are not in contact with each other. The whorls increase in diameter from the apex to the base and the broad end of the shell is called as the body whorl which has an opening called the aperture. The closed posterior end forms the protoconch. All the tapering whorls, except the body whorl, together form the spire of the shell. The angle made by the lines drawn from the protoconch to the base of the shell gives the spiral angle (Fig. 6.2 a and b).

The shape of the aperture may be oval, oblong, elongated or slit-like. The margin of the aperture is called as the peristome. The inner and the outer margin of the peristome forms the inner and the outer lip respectively.

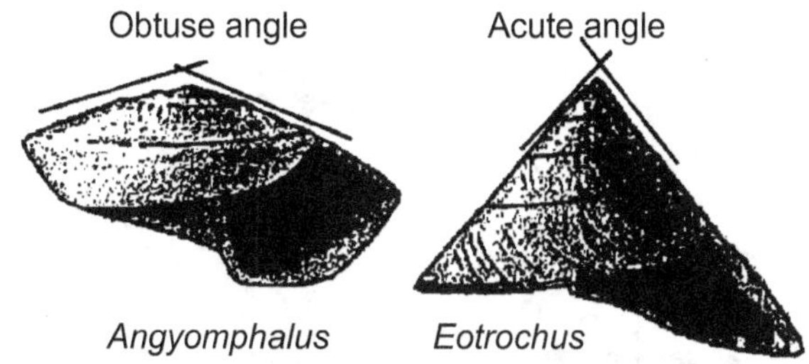

Fig. 6.2: Acute and obtuse angles of gastropod shells

In some gastropods, the peristome is unbroken and the shell is then described as holostomatous, e.g., *Turbo* (Fig. 6.3 a).

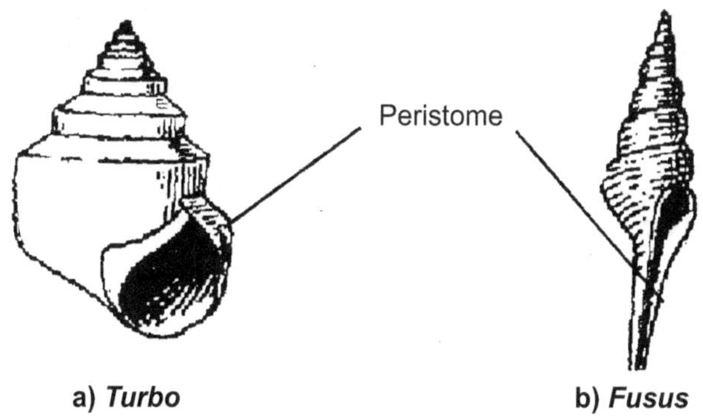

a) *Turbo*   b) *Fusus*

Fig. 6.3: Holostomatous and siphonostomatous shells

However, in many gastropods the peristome forms a canal-like extension towards the anterior end and such shells are then described as siphonostomatous e.g., *Fusus* (Fig. 6.3 b). Some gastropods like *Cyprea* of the siphonostomatous type exhibit both anterior and posterior siphonal canals.

Based on the type of coiling, gastropods exhibit either dextral or sinistral type of coiling. To illustrate this, hold the shell in such a way that the apex (posterior end) is pointed away from the observer. In this position, if the aperture is placed on the right hand side then the coiling is said to be dextral e.g., *Turbo* and *Conus* (Fig. 6.4 a), whereas the coiling is said to be sinistral, if the aperture is placed on the left hand side e.g., *Physa* (Fig. 6.4 b).

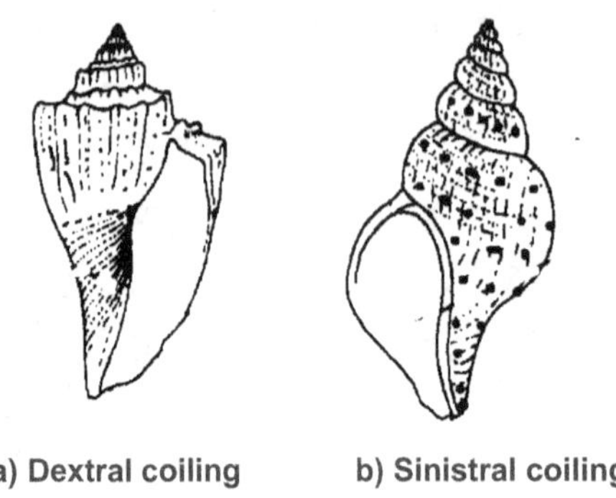

a) Dextral coiling   b) Sinistral coiling

Fig. 6.4: Types of coiling in gastropod shells

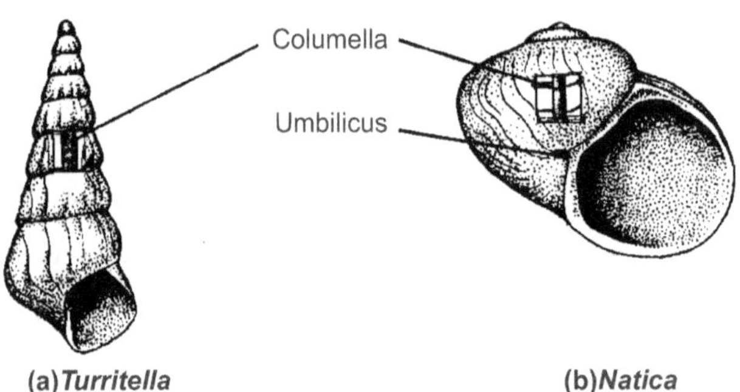

(a)*Turritella*   (b)*Natica*

Fig. 6.5: Imperforate and perforate gastropod shells

The inner parts of the whorls are often fused to form a solid axial pillar which extends from the apex to the base of the shell and is called the *Columella* (Fig. 6.5). In a cross section of the shell, the columella appears like the back bone strengthening the shell. The shells are termed imperforate when the solid columella does not open at the base e.g., *Turritella* (Fig. 6.5 a). In some gastropods, however, the inner parts of the shell are not fused but have hollow tube which opens at the base of the shell. This opening is called the umbilicus and the shells are described as perforate, e.g., *Natica* (Fig. 6.5 b).

Ornamentation

The surface ornamental features of the gastropods commonly include varying types of colouration and sculptural markings which make these shells the most sought after objects on the beaches. The ornamentation consists of features such as tubercles, knobs, spines, ribs, nodes, sculptures, granules or pits (Fig. 6.6).

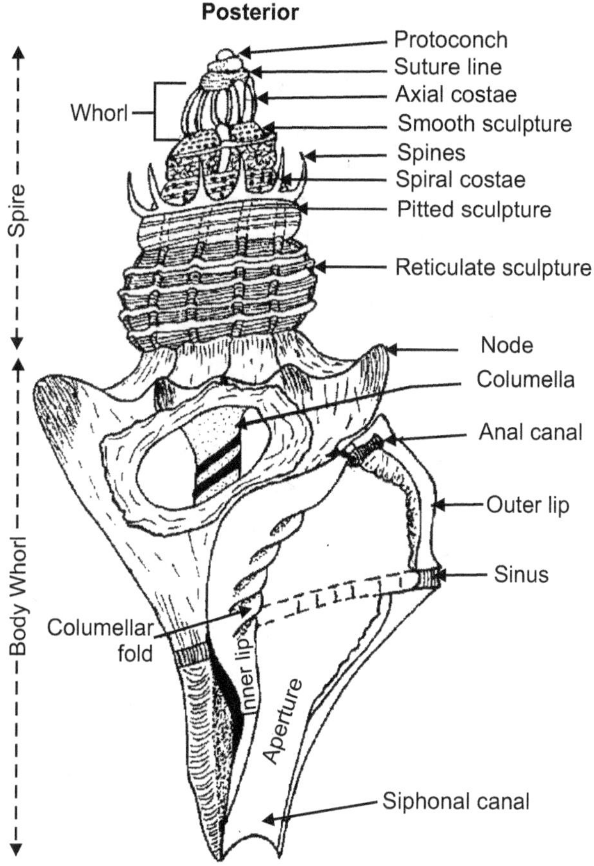

Fig. 6.6 : Ornamentation in Gastropods

These may be arranged symmetrically or separately. In some genera (e.g., Murex) rows of spines or lamellar processes, extend across all the whorls from the apex to the base of the shell, forming varices. Tubercles may be of varying sizes with finer ones all over the shell and larger ones mostly appearing in sets of two (termed as spiral or axial coastae). These features are either parallel or transverse to the sutures. In addition, the shells are beautifully and brightly coloured all over their outer surface, often variegated. In some fossils, the colour fades away due to burial for longer span of time but it is retained in few specimens, although fossils belong to the Carboniferous period. It is established that the ornamentation has evolved from simple to very complex patterns over geological time. The shell consists of an outer chitinous layer, and of a calcareous layer, usually aragonite, which is thick and porcellanous. In some cases, there is also an inner nacreous or pearly layer.

6.3 FORMS OF GASTROPODS

The forms of the gastropods are described taking into account the arrangement of the whorls, the spiral angle, the number, shape of the whorls, the size of the last whorl and its relation with the spire.

The different forms of gastropods are described below with illustrations Fig. 6.7.

(a) **Advolute:** The whorls are found to be in contact but not embracing, e.g., *Planorbis*.

(b) **Evolute:** These are loose-coiled shells where the whorls are not in contact, e.g., *Ecculiomphalus*.

(c) **Involute:** These shells have their outer whorls strongly embracing the inner whorls, e.g., *Owenella*.

(d) **Discoidal:** The shape of these shells is like a disc or wheel. The spire is extremely short and flat, e.g., *Planorbis*.

(e) **Trochiform:** The shells are conical in shape with a characteristic flat base. The sides of these shells are evenly confluent and are almost straight, e.g., *Trochus*.

(f) **Turbinate:** The shells are conical in shape with a characteristic rounded base, e.g., *Turbo*.

(g) **Turreted:** The shells have numerous whorls which increase in diameter evenly towards the base, e.g., *Turritella*.

(h) **Fusiform:** These shells are spindle shaped, thickest at the center and pointed at both the ends, e.g., *Fusus*.

(i) **Cylindrical:** The shell is roughly cylindrical in shape. The whorls are nearly equal in diameter except the first few initial whorls, e.g., *Pupilla*. (hence, also called as Pupaeform).

(j) Globular: The spire of these shells is small and short while body whorl is large and rounded, e.g., *Natica*.

(k) Biconical: The spire and the base of the body whorl together possess a conical shape, e.g., *Conus*.

(l) Platelliform: These are non-coiled shells which are low cap-shaped, e.g., *Platella*.

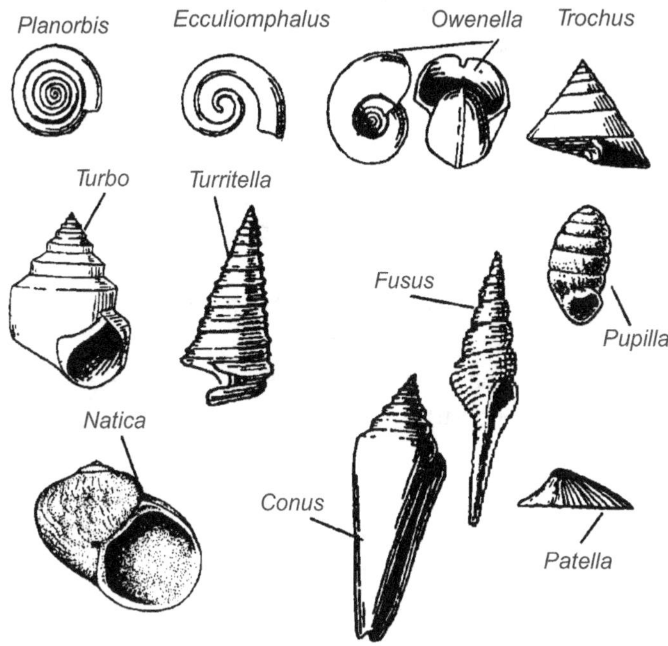

Fig. 6.7: Typical forms of gastropod shells

POINTS TO REMEMBER

- Gastropoda (in Greek *Gastros* - stomach, *Podos* - foot) is a major class of the Phylum Mollusca which includes animals which bear a coiled (e.g., *Turbo*) or uncoiled (e.g., *Patella*) calcareous shell, while others like slugs have no hard parts.

- They are found in sedimentary rocks of all ages and their presence as living animals, as well as, fossils helps the palaeontologists to unravel the geological environments which existed in the past.

- The gastropod shells are univalved as they consist of only one valve.

- The shell is coiled spirally into a screw like structure and each of the individual coils is described as a whorl.

- The forms of the gastropods are described taking into account the arrangement of the whorls, the spiral angle, the number, shape of the whorls, the size of the last whorl and its relation with the spire.

EXERCISE

1. Describe the morphology of the gastropod shell with reference to the spire and the body whorl.

2. Describe, with the help of neat diagrams, different types of coiling in the gastropods.

3. Give an account of the different forms of the gastropod shells with the help of neat diagrams and examples.

4. Write notes on
   (a) Aperture and body whorl of the gastropod shell
   (b) Coiling in gastropod shells
   (c) Orientation in gastropod shells
   (d) Forms of gastropod shells.

Chapter **7** ...

CLASS CEPHALOPODA

Contents ...

SYSTEMATIC POSITION

Kingdom	Animal
Sub-kingdom	Invertebrata
Phylum	Mollusca
Class	Cephalopoda (animals with tentacles over their head)
Subclass	Nautiloidea (Suture lines simple with lobes and saddles, e.g., nautilus)
	Subclass Ammonoidea (Suture lines complex e.g., *ammonoid*) **Tetrabranchia**
Subclass	Coleoidea (Shell is either internal or absent, presence of two gills, e.g., *octopus, belemnites*) **Diabranchia**

7.1 INTRODUCTION

Most of the molluscs described earlier are relatively slow moving, sessile invertebrates which mostly crawl or drift along with the water waves or currents in search of food or shelter. However, the cephalopods like cuttle fish and squids are equipped with highly developed eyes and sensory organs. Some possess tentacles either for swimming or catching its prey. When threatened by enemies, they may spit clouds of inky fluids.

Cephalopods are exclusively marine invertebrates which have been recorded since Cambrian. The term cephalopod has been coined from Greek terminology *cephalon* - head and *podos* - foot. Due to the restricted appearance of some of the cephalopods, they have been used as index fossils in the Palaeozoic and Mesozoic rocks. The living cephalopods include squids, nautilus and octopus. Subclasses Nautiloidea and ammonoidea have four gills for respiration, hence, the term tetrabranchia (*tetra* - four, *branchia* - gills), while Coleoidea possess only two gills and are classified as diabranchia (*dia* - two, *branchia* - gills). These have been described separately in the text to follow.

7.2 NAUTILUS

7.2.1 Morphology of Hard Parts of Nautilus

Nautiloids, today, are represented by their four genus in the marine waters of Fiji Islands in Pacific ocean. Therefore, they are referred to as living fossils. A typical nautiloid shell is about 20 cm in diameter. The shell is smooth and ornamented with radial colour banding of orange-brown stripes (Fig. 7.1).

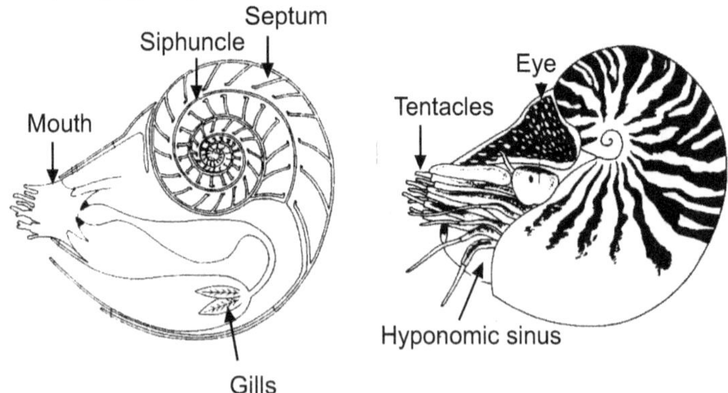

Fig. 7.1: Morphology of *nautilus* shell

The external coiled shell of Nautilus is essentially composed of calcium carbonate and an admixture of organic matter. The shell is like a hollow cone which is closed and pointed at one end while open and broad at the other end. This tube is coiled in a plane (planispiral) so that it appears like a disc and in cross-section it exhibits a horse-shoe shape. This planispiral shell possesses a bilateral symmetry. Each coil forms a whorl and these whorls do not close tightly around the axis of coiling. They leave a depression called umbilicus which may remain open or closed by a calcareous deposit termed as umbilical plug. On the basis of the size of umbilicus, nautiloids are grouped into two categories, namely, evolute nautiloids with broad and wide umbilicus and involute nautiloids with small sized slender umbilicus. The broad, living chamber of the animal has an opening in the front known as the aperture. Nautilus is able to draw its head into the shell and close the aperture with its tough hood. This aperture has a shallow embayment called hyponomic sinus which forms the ventral side while the side opposite to it is the dorsal side (Fig. 7.1).

7.2.2 Coiling in Nautiloid Shells

The earliest known nautiloids had a slightly curved shell known as cyrtocone from which it is believed that a large variety of forms developed. Important types of forms of coiling have been described and shown below (Fig. 7.2 a to i):

(a) **Cyrtocone:** Shells are slender and curved.

(b) **Orthocone:** Shells are slender and straight.

(c) **Lituiticone:** The shells that were coiled in their early stages become straight at their maturity.

(d) **Brevicone:** The shells are short and blunt.

(e) **Ascocone:** The shells which are slender and curved in their early stages become short and blunt at maturity.

(f) **Gyrocone:** The shells which are loosely coiled.

(g) **Advolute:** Coiled shells with whorls touching each other.

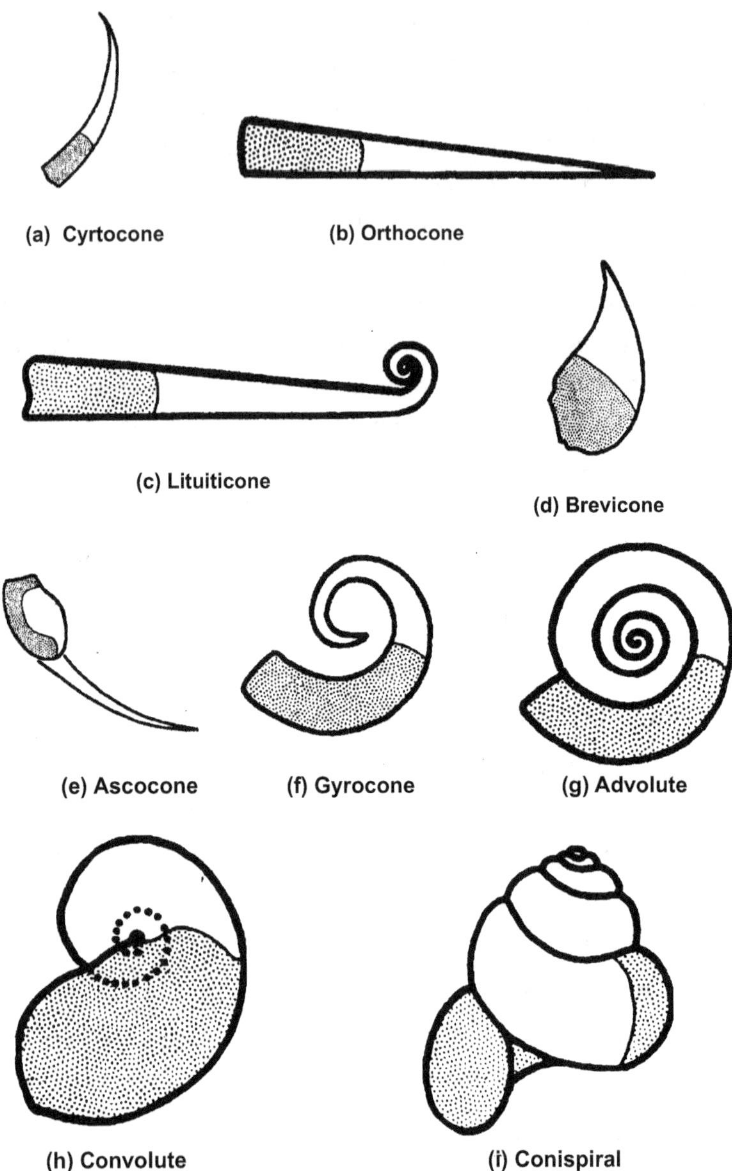

Fig. 7.2: Types of coiling in nautiloids

(h)   **Convolute:** Shells with tight coiling where the outer whorls embrace the inner ones.

(i)   **Conispiral:** Coiling of the shell takes place like a screw resembling a gastropod shell.

7.2.3 Internal Morphology

A section through the nautilus shell reveals that the shell is divided into chambers or camerae by means of thin partitions called septa. These chambers except the body chamber are also known as gas chambers. The animal regulates its buoyancy in the water column by converting its chamber fluid into a gas and vice versa. Hence, it either sinks or floats similar to a submarine. The size of the chambers increases from the tapering end (protoconch) to the broad end of the shell. An adult nautilus shell has 33 to 36 chambers which are composed of calcareous aragonite. The septa are concave towards the aperture of the shell. As the animal grows it secretes a new septum. The last chamber in which the animal stays is referred to as a body chamber. Piercing through the center of each septum is a single delicate tube-like feature known as the siphuncle. While piercing through the septa, siphuncle passes through a funnel-like opening called septal neck (Fig 7.3a). After the death of the animal, all the soft parts including siphuncle get disintegrated. However, the septal necks get preserved along with the other hard parts. In nautilus, the septal necks point away from the aperture and hence, the shell is referred to as retrosiphonate.

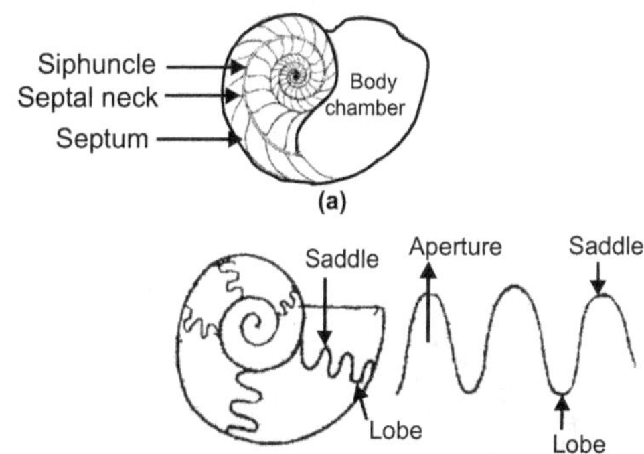

(b) Side view   (c) Orientation of saddle and lobe

Fig. 7.3: Geometry of the suture lines

When septa meet the outer wall of the shell it forms suture lines. These suture lines are very useful in classifying fossil nautiloids. In nautiloids, the suture lines are of simple type, i.e., usually straight or undulating. The convex part of the suture line towards the aperture is termed as a saddle while the concave part forms the lobe (Fig. 7.3b). These saddles and lobes have bilateral symmetry and therefore, while drawing their sketches a median arrow head is shown which points towards the aperture (Fig. 7.3c).

7.3 AMMONOIDS

7.3.1 Morphology of Hard Parts of Ammonoids

Similar to nautiloids, ammonoid shell is planispirally coiled and have complex fluted sutures. There are index fossils of the Mesozoic era. The diameter of the fossil varies from one centimeter to about three meters (e.g., *Pachydiscus*). Most of the planispirally coiled ammonoid shells exhibit all degrees of involution of the whorls. Accordingly, a shell in which all the previous whorls are exposed is called evolute and such shells have broad umbilicus. In certain shells, due to tight coiling, the last whorl covers all the previous whorls and is referred to as involute. The umbilicus of these shells is narrow as shown in Fig. 7.4.

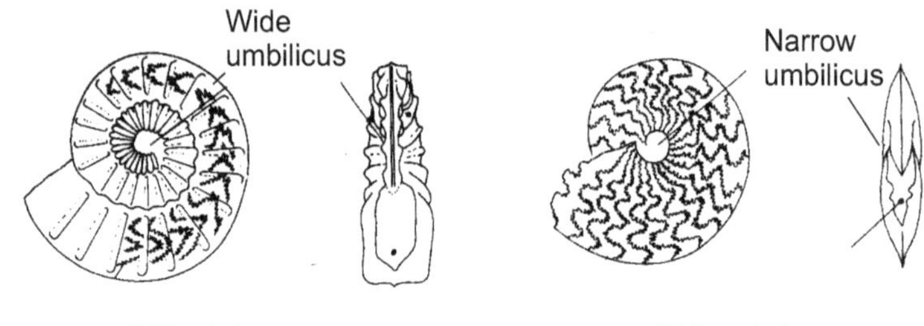

a) Evolute　　　　　　　b) Involute

Fig. 7.4: Involutions in Ammonoid shells

During certain periods in the geological history some groups of ammonoids evolved shells of highly abnormal forms, which are referred to as heteromorphs (Fig. 7.5).

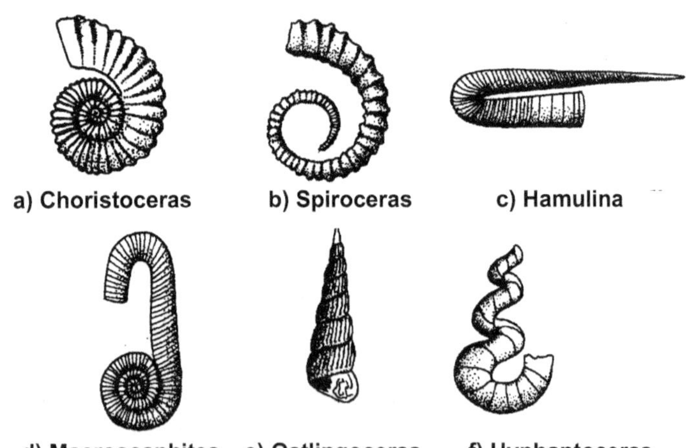

a) Choristoceras　　b) Spiroceras　　c) Hamulina

d) Macroscaphites　e) Ostlingoceras　f) Hyphantoceras

Fig. 7.5: Heteromorphs in ammonoids

Heteromorphism took place during Triassic, Jurassic and Cretaceous periods. As shown in the figure above, heteromorphic shells may have following variations.

1.　Loosely coiled shells with whorls partially or completely separated, e.g., *Choristoceras* and *Spiroceras* (Fig. 7.5 a and b).

2.　The shell may be straight, e.g., *Bochianites* or like a walking stick e.g., *Hamulina* (Fig. 7.5 c).

3.　Some shells had early coiled whorls which became straight and finally recurved near its aperture e.g., *Macroscaphites* (Fig. 7.5d).

4.　Few ammonoids had a morphology of a typical helically coiled gastropod shell e.g., *Ostlingoceras* (Fig. 7.5e).

Few ammonoids had shells which were open or loosely coiled and were asymmetrically helicoid in shape e.g., *Hyphantoceras* (Fig. 7.5 f).

In regards to their ornamentation, Palaeozoic ammonoids were unornamented. However, many Mesozoic forms were highly ornamented with growth lines, transverse ribs, protuberances ranging from gentle swellings to nodes and spines.

The last chamber in which the animal stayed is referred to as a body chamber (Fig. 7.6).

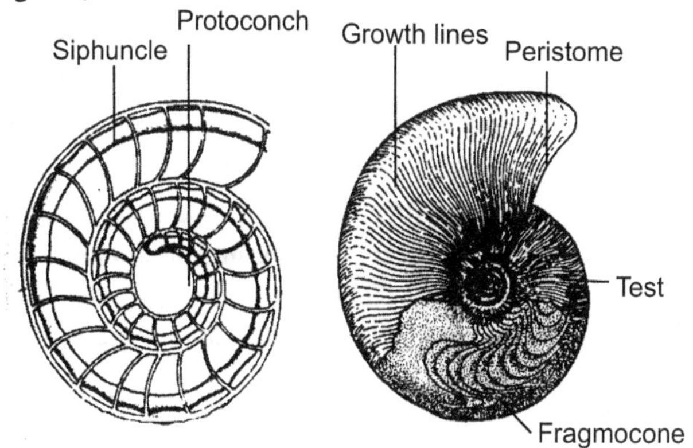

Fig. 7.6: A typical ammonoid shell

The opening of the body chamber is called an aperture. The aperture of the animal gets modified as the animal reaches its maturity. In living conditions, the aperture is covered by a single horny plate, the anaptychus.

Transverse section of an ammonoid reveals the same morphology as that of the nautilus shell. The shell is divided into numerous chambers by means of septa. The size of the septa increases from protoconch towards the aperture of the shell. These chambers are believed to be used for regulating the buoyancy of the animal. Their function, however, is similar to that in nautilus. These septa are slightly curved in the center but it becomes frilled towards the outer wall, where it becomes the suture. Piercing through the septa is the siphuncle. In most ammonoids, the siphuncle is situated near the outer margin (ventral). The extension of the setpa forms the septal necks which appear as funnel shaped structures which are directed towards the aperture of the shell and hence, termed as prosiphonate.

7.3.2 Types of Suture Lines

When septa meets the outer wall of the shell it forms suture lines. In ammonoids, the suture lines are of complex nature. Hence, depending upon the nature of these suture lines, ammonoids exhibit following types of suture lines as shown in Fig. 7.7 (a to c).

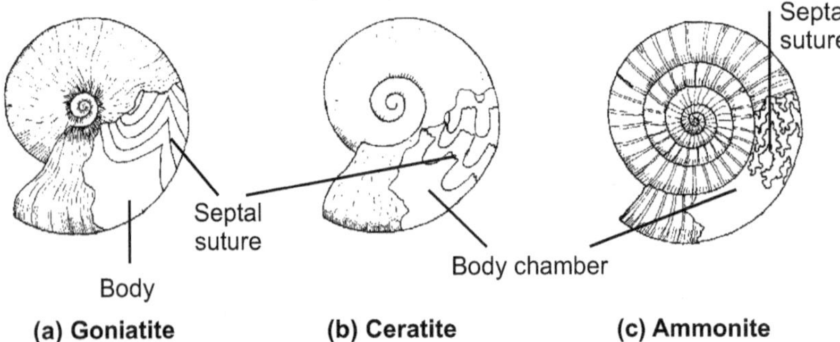

Fig. 7.7: Types of suture lines in ammonoids

(a) Goniatite suture: The Palaeozoic ammonoids were characterised by sharp angular lobes and rounded saddles without any accessory crenulations (Fig. 7.7a).

(b) Ceratite suture: The Triassic ammonoids are characterised by frilled lobes and rounded saddles (Fig. 7.7b).

(c) Ammonite suture: The Mesozoic ammonites have finely subdivided complex lobes and saddles. However, some Permian ammonoids with this type of suture lines have also been recorded (Fig. 7.7c).

The suture lines in Ammonoids are of immense importance in their classification and identification.

7.4 BELEMNITES

7.4.1 Morphology of Belemnites

The subclass Coleoidea is represented by modern squids or cuttlefish. However, the sediments of Jurassic and Cretaceous age are often found to contain abundance of belemnites. These fossils are characterised by three parts namely, the guard, phragmocone and the pro-ostracum (Fig. 7.8).

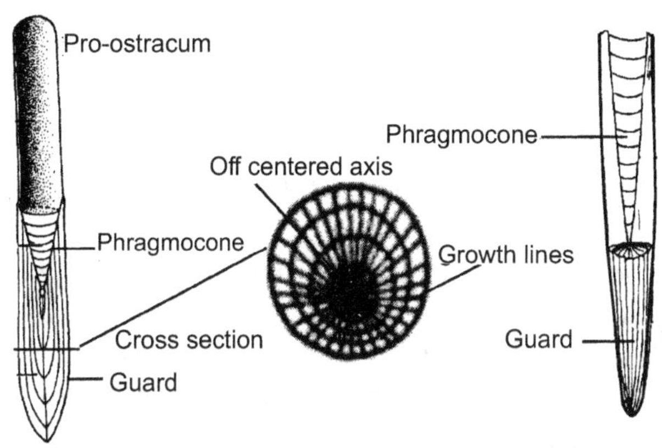

Fig. 7.8: Morphology of belemnite fossil

In majority of cases, all these parts are not preserved in toto (together). Guard of the belemnite is the most abundant fossil preserved in sediments. The other two parts get eroded or disintegrated due to their susceptibility to erosive agents. However, their preservation is not uncommon. It is, therefore, convenient to describe these parts separately.

Guard: The posterior part of a belemnite consists of a hard, bullet-shaped, cylindrical solid made up of calcite. It is known as the 'guard' of the belemnite as shown in Fig. 7.8 a. The sides are often found to be parallel, tapering posteriorly to a conical apex. The anterior part of this guard encloses a deep 'V' shaped cavity called as alveolus into which the phragmocone is embedded. The length of the guard varies from one to twelve inches. The guard helps the animal to maintain its horizontality inside the water column.

Transverse polished section of the guard reveals that it is made up of radiating fibers of calcite. In addition to this radial calcitic structure, growth layers resembling somewhat like the growth lines of a tree trunk are observed (Fig. 7.8a). This cross section also reveals that its axis of the radiating calcitic fibers forms the apical line, which is slightly off-centered.

Phragmocone: The conical, thin walled chambered part of the shell which fits into the alveolus of the guard is known as the phragmocone. It is aragonitic in composition (Fig. 7.8 c). It normally projects outside the alveolus. The phragmocone is divided into number of chambers by means of septa which are concave anteriorly. A slender siphuncle pierces through the septa at the ventral margin. The lower part of the phragmocone bears a tiny bulbous protoconch.

Pro-ostracum: The dorsal part of the phragmocone extends far beyond the last-formed septum into a leaf-like or a blade-like projection called pro-ostracum (Fig. 7.8 a). This projection is believed to be the rudimentary living chamber. This part being very delicate is rarely preserved as fossils.

It is believed that the tentacles of the animal had horny hooks which sometimes are preserved as fossils. For example, in Europe, Jurassic rocks contain some well preserved belemnites with arm hooks.

Table. 7.1: Comparison between Nautiloidea and Ammonoidea

Order Nautiloidea	Order Ammonoidea
1. Septa are simple or slightly concave.	1. Septa are complexly folded.
2. Suture line is simple.	2. Suture line is complex.
3. Siphuncle is centrally located.	3. Marginal siphuncle (placed at outer margin).
4. Retrosiphonate septal necks.	4. Prosiphonate septal necks.
5. These are generally coiled, (orthoceras is an uncoiled example).	5. These are also coiled, but earlier forms are uncoiled.
6. Represented today by four species.	6. They are extinct.
7. Nautilus evolved during Upper Cambrian period.	7. Ammonoids evolved from Nautiloids during Devonian.
8. They were dominant during Silurian period.	9. They were dominant during complete Mesozoic era.

Points to Remember

- Cephalopods are exclusively marine invertebrates which have been recorded since Cambrian.
- The living cephalopods include squids, nautilus and octopus.
- Subclasses Nautiloidea and ammonoidea have four gills for respiration.
- Coleoidea possess only two gills.
- Nautiloids are represented by their four genus in the marine waters.
- The earliest known nautiloids had a slightly curved shell known as cyrtocone.
- The nautiloids are typically shallow water, marine invertebrates.
- The ammonoid shell is planispirally coiled and have complex fluted sutures.
- They are the index fossils of the Mesozoic era.
- Belemnites are exclusively marine animals revealing a typical necto-benthic habitat.
- Belemnites are characterised by three parts namely, the guard, phragmocone and the pro-ostracum.
- The posterior part consists of a hard, bullet-shaped, cylindrical solid made up of calcite. It is known as the 'guard' of the belemnite.
- The conical, thin walled chambered part of the shell which fits into the alveolus of the guard is known as the phragmocone.
- The dorsal part of the phragmocone that extends far beyond the last-formed septum into a leaf-like or a blade-like projection is called pro-ostracum.

Exercise

1. Describe the morphology of hard parts of Nautiloids.
2. Describe the morphology of hard parts of Ammonoids.
3. Describe the morphology of hard parts of Belemnites.
4. Distinguish between the Nautilus and Ammonoids.
5. Write notes on
   (a) Types of coiling in Nautilus
   (b) Types of suture lines in Cephalopods
   (c) Guard and Phragmocone of Belemnites

Chapter **8**...

PHYLUM BRACHIOPODA

Contents ...

SYSTEMATIC POSITION

Kingdom	Animal
Sub-kingdom	Invertebrata
Phylum	Brachiopoda
Class	Articulata (valves are hinged by teeth and socket arrangement)
(1) Order	Palaeotremanta (hinge is simple but lacks teeth and socket arrangement) e.g., *Rustella*
(2) Order	Protremata (presence of pseudo deltidium only in the younger stages) e.g., *Productus*
(3) Order	Telotremata (highly modified pedicle opening and brachial skeleton) e.g., *Spirifer, Terebratula*
Class	Inarticulata (valves are joined by complex set of muscles).
(1) Order	Atremata (lack of pedicle opening) e.g., *Lingula*
(2) Order	Neotremata (pedicle opening is present only in the younger stages) e.g., *Crania*

8.1 INTRODUCTION

Brachiopods are marine invertebrates mostly confined to the shallower depths of the sea, i.e., the area between the shore line and 200 meters depth. The Brachiopods are always fixed to the substratum and hence, they are called as sessile and / or benthos animals. However, they are free swimmers during the larval stages. The name brachiopoda was coined by the earlier workers who presumed that the presence of brachia or lophophore was to facilitate locomotion (In Greek, *brachio* - arm; *podos* - foot).

8.2 MORPHOLOGY OF THE HARD PARTS

The soft parts of the brachiopods are enclosed by two valves which have a chitinophosphatic or sometimes a calcareous composition. Brachiopod shells are easily distinguished from lamellibranch shells by the unequal valves and their position of attachment to the substratum.

The side towards the pedicle opening is the posterior side, while the side opposite to it, is the anterior side. Of the two valves, one is dorsal and the other is ventral in position with regard to the body parts of the animal (Fig. 8.1a). The two valves meet along a curved line referred to as a commissure. The plane drawn through the commissure is horizontal and hence is called the commissural plane (Fig. 8.1b). In the brachiopods, the plane of symmetry is median and vertical (Fig. 8.1c). In addition, all brachiopod shells have an equilateral symmetry. Most of the living brachiopods have a size of a few mm upto 80 mm but there are some fossil brachiopods which have attained a width as much as 370 mm and a length of 250 mm.

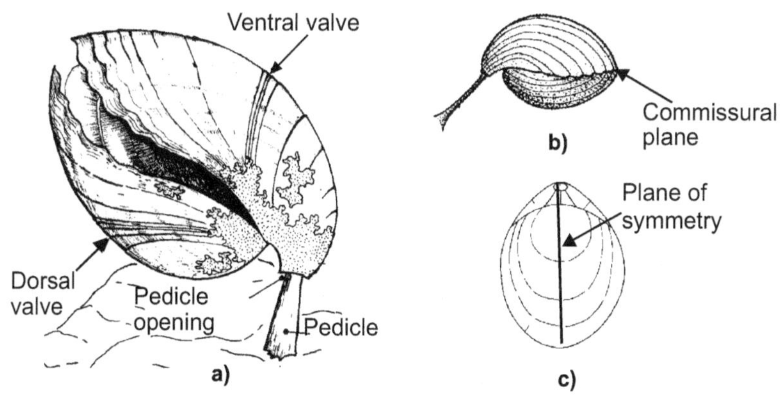

Fig. 8.1: Position and symmetry of Brachiopod valves

The dorsal valve is much smaller in size when compared to the ventral valve which gives rise to an inequivalved shell. At the time of the attachment to the substratum, the animal reverts its position so that the dorsal valve goes down as shown in the figure 8.1a. The ventral valve is also referred to as the pedicle valve as the pedicle emerges through an opening known as the foramen. The pedicle is a muscular stalk which helps the animal to attach itself to the substratum. However, it is destroyed on the death of the animal and only the foramen remains on the fossil shell. In some forms of brachiopods, the pedicle is absent and such forms are instead cemented to the soft sediments by means of spines. Both the ventral and dorsal valves show a beak-like structure at their posterior margin which is called the umbo.

In Fig. 8.2a, the distance X-Y measures the width of the shell while P-Q, the distance between the umbo and the anterior margin of the shell, is its length. In Fig. 8.2 b, the distance M-N between the attached valves gives the thickness of the shell.

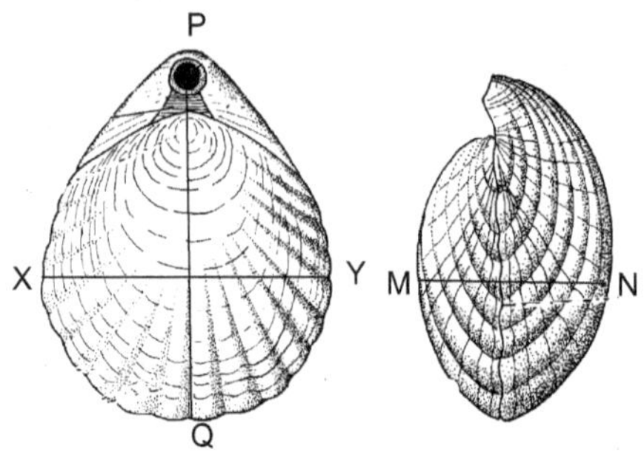

Fig. 8.2: Thickness, length and width of a Brachiopod shell

Between the umbo and the inner edge of the valve there is a triangular cavity called the delthyrium which is closed by a pair of deltidial plates. The two valves are joined along a line commonly referred to as the hinge line. The hinge line may be straight as in *Spirifer* or curved as in *Terebratula*. The exterior surface of the valves may be smooth and marked with concentric growth lines or covered by spines or by radial ridges.

8.3 INTERNAL MORPHOLOGY

The inside of the valves shows a complex system of muscles which are joined to the shell and are used by the animal for its various functions. With the death of the animal, the muscles and the soft parts decay leaving behind only the scars. Internally, the pedicle valves show three sets of muscle scars. Of these, the first set of scars are called adductor muscles which were once used for closing the shell; the diductor muscles for opening the shell; and the adjustor muscles for the movement of the shell on the pedicle (Fig.8.3a).

Apart from these muscle scars, there is a set of teeth which fit into the corresponding sockets of the other valve. Thus, the tooth of the pedicle valve fits into the socket of the brachial valve.

The brachial valve shows a pair of adductor muscles and a spirally enrolled and multifold structure which is commonly referred to as brachia (Fig.8.3b). Each of the brachia has a groove which leads back to the mouth. In the living animal, these grooves are lined by fine cilia which vibrate and thereby set up water currents which facilitate movement of food particles into the mouth.

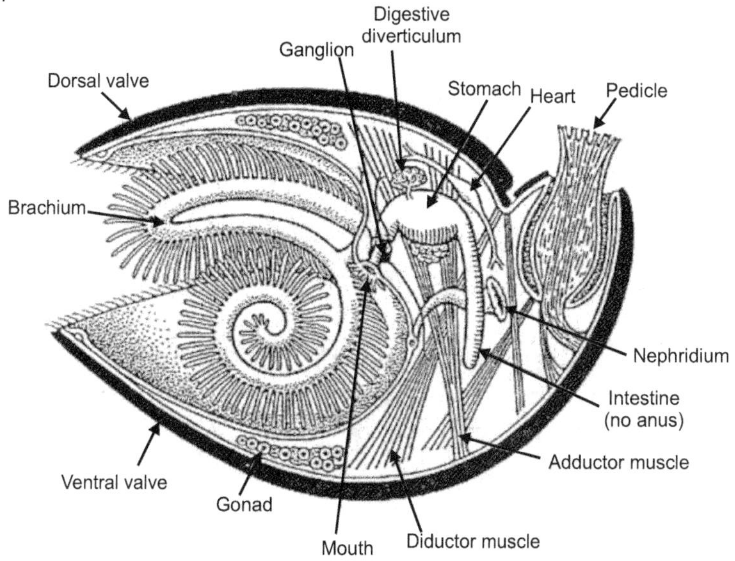

Fig. 8.3: Pedicle and brachial valves

The framework being calcareous is preserved under suitable fossilisation conditions and gives rise to the brachial skeleton. The form and shape of the brachial skeletons vary significantly in different genera as shown in Fig. 8.4 (a to g).

The following are some important types of brachial skeletons

- **(a)** **Rhynchonella:** It is a simple type of brachial skeleton which has two very small, short but curved plates known as crura, e.g., *Rhynchonella*.

- **(b)** **Terebratula:** It consists of short ribbon-like bands coming out from the crura and forming a small loop, e.g., *Terebratula*.

- **(c)** **Stringocephallus:** The skeleton ribbon is more extensive forming a large loop which runs parallel to the anterior margin, e.g., *Stringocephallus*.

- **(d)** **Magellania:** In this type the brachial ribbon extends nearly to the anterior margin of the shell and then bends back on itself twice, e.g., *Magellania*.

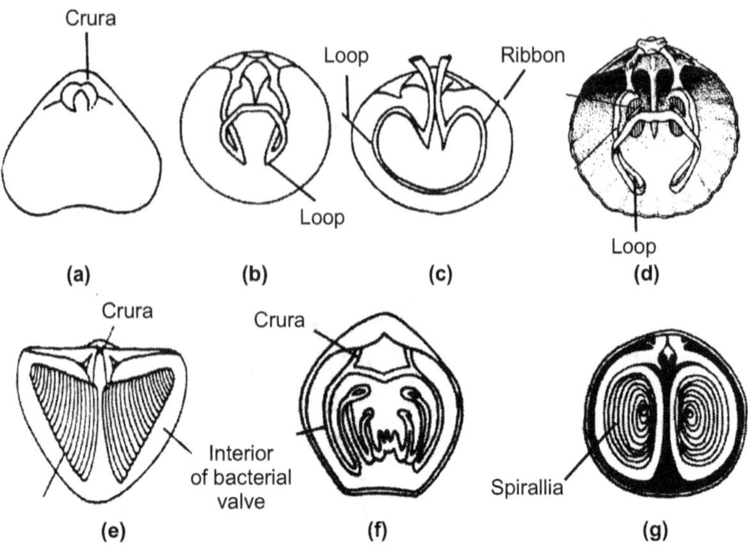

Fig. 8.4: Forms of the Brachial skeleton

(e)　**Spirifer:** The brachial skeleton forms two spiral ribbons arising from crura. These coils or spiralia point towards the lateral margins, e.g., *Spirifer*.

(f)　**Glassia:** The brachial skeleton forms spiral ribbons unlike in spirifer. Its coils are turned up and down with the apices pointing inward, e.g., *Glassia*.

(g)　**Atryopa:** The spiral ribbon arises from the crura, forming spirals with their apices pointing towards its posterior end, e.g., *Atrypa*.

8.5　COMPARISON BETWEEN THE BRACHIOPODS AND LAMELLIBRANCHS

Brachiopods	Lamellibranchs
1.　Brachiopods are exclusively marine.	1.　Lamellibranchs are found in both marine as well as fresh water conditions.
2.　The shell is bivalved and valves are left and right in position.	2.　Shell is bivalved and are left and right in position.
3.　They are sessile and benthos.	3.　With some exceptions, majority of them can move freely.
4.　The shell is equilateral.	4.　The shell is inequilateral.

Contd...

5. Inequivalved nature of the shell.	5. Equivalved nature of the shell.
6. The median plane is horizontal.	6. The median plane is vertical.
7. The shell is perforated with an opening in the ventral valve.	7. Perforation absent.
8. The brachial skeleton is present.	8. The brachial skeleton is absent.
9. The hinge line marks the posterior dorsal side.	9. The hinge line marks side.
10. The hinge line shows two short curved teeth in the ventral valve and corresponding sockets in the dorsal valve.	10. The hinge line shows an equal distribution of teeth and socket in both valves.
11. There are three sets of different muscle scars (adductor, diductor and adjustor).	11. There is only one set of adductor muscles.
13. The shell opens anteriorly.	13. The shell opens ventrally.
14. The ventral valve is more prominent.	14. Both valves are of equal size.
15. The shell is chitinophosphatic and also calcareous in composition.	15. The shell is mostly aragonitic in composition.
16. Brachiopods are less numerous in the present day marine waters.	16. They are present in abundance today.

POINTS TO REMEMBER

- The soft parts of the brachiopods are enclosed by two valves which have a chitinophosphatic or sometimes a calcareous composition.

- The two valves meet along a curved line referred to as a commissure.

- The dorsal valve is much smaller in size when compared to the ventral valve which gives rise to an inequivalved shell.

- Between the umbo and the inner edge of the valve there is a triangular cavity called the delthyrium which is closed by a pair of deltidial plates.

- The inside of the valves shows a complex system of muscles which are joined to the shell and are used by the animal for its various functions.

- The brachial valve shows a pair of adductor muscles and a spirally enrolled and multifold structure which is commonly referred to as brachia.
- Each of the brachia has a groove which leads back to the mouth.
- The important types of brachial skeletons are Rhynchonella, Terebratula, Stringocephallus, Magellania Spirifer, Glassia, Atryopa.
- The Brachiopods are mostly clustered in the marine environment of the tropical regions.

EXERCISE

1. Describe the morphology of the hard parts of the Brachiopod shell with the help of neat diagrams.
2. Differentiate between the Brachiopod and Lamellibranch shells.
3. Write notes on:
   (a) Classification of Brachiopods.
   (b) Brachial skeleton in Brachiopods.
   (c) Articulate and inarticulate Brachiopods.
   (d) Symmetry of Brachiopod shell.
   (e) Pedicle and brachial valves of Brachiopods.

Chapter **9**...

PHYLUM COELENTERATA

Contents ...

SYSTEMATIC POSITION

Kingdom	Animalia
Sub-kingdom	Invertebrata
Phylum	Coelenterata (animals possess gastro-vascular cavity) or Cnidara (Simplest of the multicellular organisms)
Class	Anthozoa (flower like animals)
Sub-Class	Zoantharia

9.1 INTRODUCTION

The animals which possess a gastro-vascular cavity or coelenteron have been assigned the status of a phylum called Coelenterata (In Greek, *Koilos* - hollow, *enteron* - intestine). The living coelenterates in the modern oceans include hydra, jelly fish, sea anemones and corals. The extinct forms are found as fossils and they include *Calceola* and *Montlivaltia*. Madreporarian shells represent an extensive order of Anthozoa including most species that produce stony corals. These usually form colonies and always have an ectodermal calcareous skeleton

Of these, the corals are exclusively marine benthic organisms and occur in the tropical climatic regions of the earth. They thrive in warm waters having a temperature range between 18° to 28° C and shallow depths up to 200 meters. They develop either as solitary or colonial forms. Both these forms have an exoskeleton covering and the body exhibits a radial symmetry. The temperature and depth factors which govern the existence of corals make these organisms of great valve to the palaeontologists. Some of the well known coral colonies of the earth include the coral reefs of the Tasman Sea (Australia), the Andaman and Nicobar islands (India) and the Bahamas (N. America). These examples have been studied in great detail and they helped in understanding the nature and the behaviour of these animals to unfold the palaeoclimatic conditions.

9.2 MORPHOLOGY OF HARD PARTS

The corals usually have two main body forms viz.

(1) The polyp which is attached to the substratum and

(2) The medusa which is a free swimming form

Madrepora meaning mother of pores, is a genus of stony corals, that develop reefs or islands in tropical locations. The names Madrepore and Madreporaria were formerly applied universally to any stony coral of the family Scleractinia. The animals belonging to the class Anthozoa do not produce medusa and hence, the description of only the polyp forms is given in the text to follow. A simple polyp has a conical or cylindrical shape and it is attached to the substratum with its tapering end. The other end, however, with tentacles and mouth remains open. The mouth not only takes in the food particles, but also serves as an anus which excretes the undigested food.

The body wall of the polyp consists of the outer layer of cells called the ectoderm and an inner layer, the endoderm. Between these layers, there is a jelly-like layer of mesogloea having a nervous system. The cross section of the polyp is shown in Fig. 9.1.

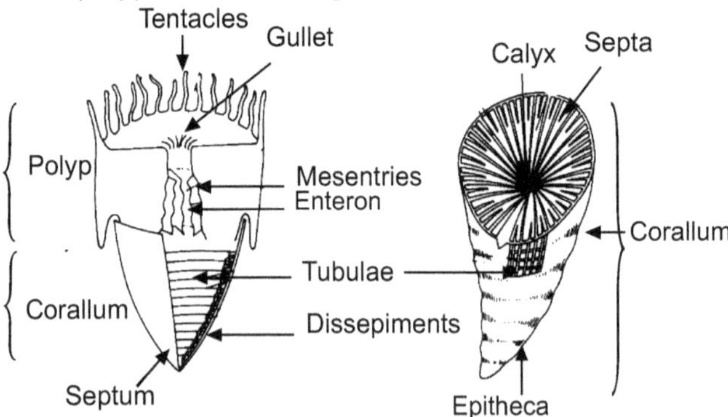

Fig. 9.1: Structure of an individual polyp

The mouth of the polyp leads to the body cavity called enteron. The terminal end of the polyp has a cup shaped, open depression called as the calyx. The calcium carbonate skeletons of the living corals are usually aragonitic which later on recrystallises into calcite on fossilization of the dead coral. The exoskeleton of an individual animal is called as a corallite and as corallum, when it develops into a colony. The outer wall of the coral is called as the theca, while the calcite coating on it, gives rise to the epitheca. The epitheca shows a number of growth bands on its surface which gives it a rough appearance (rugose texture). The growth bands are secreted by the polyp. The corallum in different species may form either compacted or separate tubes. Hence, the morphology of the corallites gives rise to different simple and compound forms like horn corals (like a cone), button corals (like a disc), brain-corals (like a dome), branching corals and picket-fence corals (Fig. 9.2).

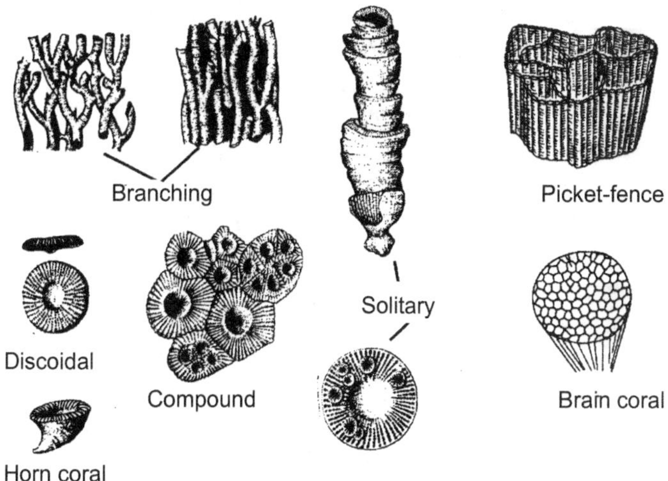

Fig. 9.2: Different types of corals

The food is passed on into the enteron with the help of tentacles through the mouth. To facilitate the assimilation of the food by the endoderm, the wall of the endoderm is thrown into a series of folds called as mesenteries.

To support the mesenteries, the interior of the corallite has several radially arranged, vertical, wall-like compartments called as septa (singular septum). The septa radiate out from the pillar of calcite called columella. Based on the extension of the septum from the inner wall of the epitheca towards the endoderm, septa are grouped into categories as primary, secondary and tertiary (Fig. 9.3).

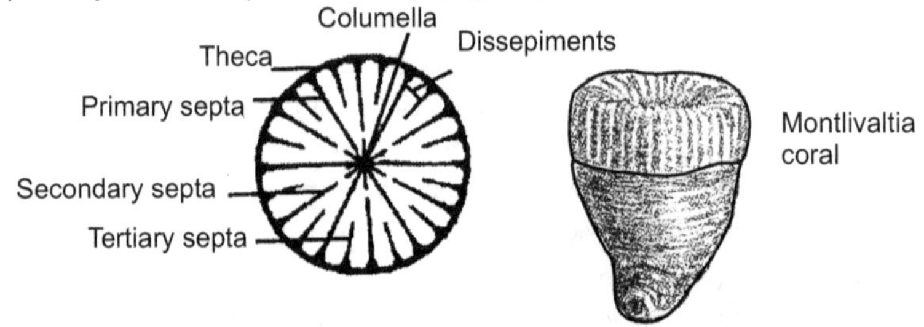

Fig. 9.3: Primary, secondary and tertiary septa

The primary septa are the longest and they reach the columella situated at the center of the corallite. The secondary and tertiary septa, however, are shorter than the primary septa as shown in the figure above. Along with the septa, there are transverse and concave rod-like plates called as dissepiments (Fig. 9.3) which help in fixing the polyp to the calyx. Some corals also possess transverse cylindrical rods known as tabulae, placed transversely to the septa. In the case of a few corals, the central columella branches off into a web-like structure called as the axial complex. The corallites which do not develop septa at a few places, create gaps which are called as fossula. The symmetrical arrangement of the septa gives the radial symmetry to the corallite. For example, in *Montlivaltia*, the number of septa varies from six or multiples of six; and in *Calceola*, there are four or multiples of four such symmetrical septa arranged on the corallite. Hence, the corals are further classified on the basis of their septal symmetry.

POINTS TO REMEMBER

- The animals which possess a gastro-vascular cavity or coelenteron have been assigned the status of a phylum called Coelenterata.
- The living coelenterates in the modern oceans include hydra, jelly fish, sea anemones and corals.
- The corals usually have two main body forms viz. the polyp which is attached to the substratum and the medusa which is a free swimming form.
- A simple polyp has a conical or cylindrical shape and it is attached to the substratum with its tapering end.

- The body wall of the polyp consists of the outer layer of cells called the ectoderm and an inner layer, the endoderm.
- Between these layers, there is a jelly-like layer of mesogloea having a nervous system.
- The mouth of the polyp leads to the body cavity called enteron.
- The terminal end of the polyp has a cup shaped, open depression called as the calyx.
- The exoskeleton of an individual animal is called as a corallite and as corallum, when it develops into a colony.
- The corals are known to be very sensitive to the temperature and depth at which they live in the sea.

EXERCISE

1. Describe the morphology of hard parts of the corals with special reference to
   (a) external morphology and
   (b) internal morphology
2. Write notes on
   (a) Septa in corals
   (b) Polyp and medusa

Chapter 10

CONCEPTS OF ORGANIC EVOLUTION

Contents ...

10.1 DEFINITION OF EVOLUTION

Organic evolution is the change in the inherited characteristics of plants and animals over successive generations. In simple terms, it states that the animals and plants have changed over long period of time and the present-day 'new' species have arisen from their pre-existing ancestors. For example, in the case of five fingered vertebrates, the limb has been modified in a number of ways. The hoofs of horses, wings of birds, flippers of marine animals and the grasping hand of primates, etc., are all the resultant of variations from a common stalk.

Such evolutionary processes give rise to diversity at every level of living forms, which include species, individual organisms and also molecular levels. Charles Darwin proposed that such an evolution takes place by means of natural selection.

Earlier, Lamarck had stated that all living organisms had originated from their ancestors and while doing so, they had slowly adapted for living in the available environments. Hence, the concept of adaptation was put forth by Lamarck. According to him, adaptation was possible only because of an existence of 'internal driving force' which made animals become more complex. He speculated that new organs had arisen due to the 'new needs' and these 'acquired characters' were inherited in the next generations.

The concept of organic evolution is based on the fact that modern species of animals and plants have originated from more primitive ancestors which were very simple life forms. Evolution, therefore, may also be defined as a change in gene frequencies through time. In any gene pool, new variations of character are added mainly due to mutation. Although, it is a slow but continuous process, palaeontologists have been able to systematically study the evolutionary trends of some of the fossils.

Lamarck stated that an organism can pass on characteristics that it acquired during its lifetime to its offspring (heritability of acquired characteristics or soft inheritance).

When Charles Darwin published his theory of evolution by natural selection 'On the Origin of Species', he continued to give credence to what he called 'use and disuse inheritance', but rejected other aspects of Lamarck's theories. Later, Mendelian genetics supplanted the notion of inheritance of acquired traits, eventually leading to the development of the modern evolutionary synthesis, and the general abandonment of the Lamarckian theory of evolution in biology. Despite this abandonment, interest in Lamarckism has continued till date, as studies in the field of epigenetics have highlighted the possible inheritance of behavioral traits acquired by the previous generation.

10.2 EVIDENCE OF EVOLUTION

Evolution by natural selection is a process inferred from three facts about populations: 1) more offsprings are produced than can possibly survive, 2) traits vary among individuals, leading to different rates of survival and reproduction, and 3) trait differences are heritable. Thus, when members of a population die they are replaced by the progeny of parents better adapted to survive and reproduce in the environment in which natural selection takes place. This process creates and preserves traits that are seemingly fitted for the functional roles they perform. Natural selection is the only known cause of adaptation, but not the only known cause of evolution. Other, non-adaptive causes of evolution include mutation and genetic drift.

10.3 MACRO- AND MICRO-EVOLUTION

The evolution may be either a micro-evolution or macro-evolution. The micro-evolution takes place within species with minor taxonomic transitions, as a result of changing interactions between the animals and the environment.

The macro-evolution, on the other hand, refers to evolution on higher levels. It is with the help of macro-evolution that scientists have been able to study the evolutionary aspects such as the origin of new structural plants and higher species of plants and animals.

Hence, the individuals are main units of selection at the micro-evolutionary level and the macro-evolution involves species at higher levels and operates over longer time periods.

Evidences of evolution are readily apparent on a micro-evolutionary level. These include observed changes in domestic crops (creating a variety of maize with greater resistance to disease), bacterial strains (development of strains with resistance to antibiotics), laboratory animals (structural changes in fruit flies), and flora and fauna in the wild (change in particular populations of peppered moths and polyploidy in plants).

It was Charles Darwin, however, in the 'Origin of Species', who first provided considerable evidences for the 'theory of descent' with modification on the macro-evolutionary level (i.e., paleontology, biogeography, morphology, and embryology). Many of these areas continue to provide the most convincing proofs of descent with modification even today. Stephen Jay Gould (1983) suggested that evolutionary modification actually comes from the observation of imperfections of nature, rather than perfect adaptations. According to him, 'All of the classical arguments for evolution are fundamentally arguments for imperfections that reflect history. They fit the pattern of observing that the leg of reptile is not the best for walking, because it evolved from fish. It means, because it is inherited from a common ancestor, that bones have a same structure, but used for different purposes, so that a rat runs, a bat flies, a fish swims or a man types.

10.3 DARWINISM AND LAMARCKISM

Darwinism is a theory of biological evolution proposed by Charles Darwin (1809–1882) and others, which states that all species of organisms arise and develop through the natural selection of small, inherited variations that increase the individual's ability to compete, survive, and reproduce. Hence, it is also known as 'Theory of Natural Selection'. Darwin, opined that out of the vast number of individuals which compete for their survival, only those having advantageous variations survive and reproduce (Fig. 10.1).

Fig. 10.1: Process of Natural Selection

This pioneering theory initiated the concept of evolution when Darwin published his views in 'On the Origin of Species' in 1859. Thomas Henry Huxley, later on, coined the term Darwinism in April 1860.

Darwinism refers to the specific concept of natural selection which is used mainly for the biological evolution. It was used to describe evolutionary concepts, which are based on the following observations:

1. **Species multiply:** in a specific geometric ratio (e.g., in certain species, population doubles itself on one year and increases four times next year).

2. **Struggle for existence:** Every species tries to keep its own population constant. During their life time, organisms compete with one another for food and shelter. During this struggle, there is a 'survival of the fittest'.

3. **Variation and heredity:** Variation includes difference between parents and off springs in their shape, size, behavior, etc. Heredity refers to the passing of characteristics through genes from one generation to the other. It eventually contributes to evolution.

With advent of time, during the late 19th century, newer scientific data were incorporated and the term 'neo-Darwinism' was coined for various alternative evolutionary mechanisms. The development of the modern synthesis in the early 20th century, incorporating natural selection with population genetics and Mendelian genetics, revived Darwinism into its modern version. Therefore, although the term *Darwinism* has remained in use, the modern evolutionary theory includes concepts such as genetic drift.

Criticism of Darwinism:

1. It does not explain the origin of variation.
2. It is not always the case that useful variations are selected.
3. Natural selection is not the only cause of specification.
4. It does not explain the presence of vestigial organs found in organisms (e.g., monkey bone in man).

Lamarckism:

Lamarckism is the hypothesis that an organism can pass on characteristics that it has acquired through use or disuse during its lifetime to its offspring. It is also known as the inheritance of acquired characteristics or soft inheritance (Fig. 10.2).

Fig. 10.2 : Evidence for Lamarckism

Although Charles Darwin is often called the father of evolution, it was **Jean-Baptiste Lamarck**, who developed and presented a logical, coherent explanation for the process of evolution. Although some of his ideas were later proven to be incorrect, his work is still really important because it laid the foundation for our modern understanding of evolutionary biology.

Lamarck believed that as a giraffe repeatedly stretched its neck, while trying to reach leaves that were higher and higher, and eventually its neck would get a little bit longer. Hence, animals, like the giraffe, actually changed as they used parts of their body in different ways. Other species also have done similar developments in the past and probably, would do in future, e.g., a polar bear would develop thicker fur if the climate suddenly gets much colder, or a duck could develop webbed feet to swim more efficiently.

He also believed that these changes would be passed on to subsequent generations. A giraffe, whose neck had been stretched longer, would have a baby giraffe whose neck would already be longer than that of the other giraffes who hadn't stretched their necks as much. Lamarck's theory, known as **Lamarckism**, is also commonly known as the theory of **inheritance of acquired characteristics** because he believed that traits that were acquired during an animal's life would be passed down to the next generation.

10.5 MUTATION

A mutation occurs when a DNA gene is changed in such a way as to alter the genetic message carried by that gene. A **mutagen** is a physical or chemical agent that changes the genetic material, usually DNA of an organism and thus, increases the frequency of mutations above the natural background level.

Heredity results accurate reproduction of genes. During cell-division, the chromosomes divide to give rise to daughter chromosomes. The chromosomes are beset with genes and consequently chromosome duplication means, the duplication of genes. Genes always arise from existing genes although the material for the synthesis comes from nutritional sources. The newly formed gene is an exact replica of the parent gene.

The process of gene reproduction, though exact, sometimes results into an error in copying. The copy of the gene, therefore, differs from the original and the modified gene reproduces its changed structure. This error in copying is mutation. Following figure explains the process of mutation in a human eye.

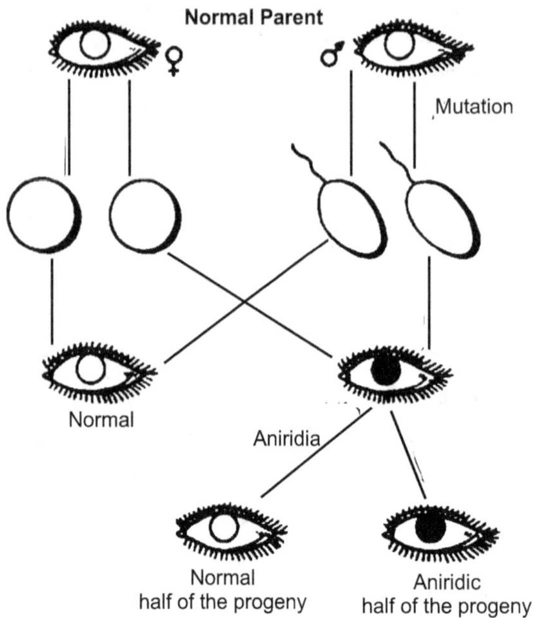

Fig. 10.3: Mutational changes in iris yielding aniridic condition

Genes are responsible for characters. In man, the gene regulates the formation of normal iris of the eye. Sometimes this gene becomes so altered that the iris fails to develop. Failure of the iris to develop is known as aniridia (Fig. 10.3). Mutation of this gene occurs in the germ cell of a man and is transferred to the zygote which finally develops into a new man.

The mutated gene is a dominant gene and as a result the new individual is born with no iris in the eye. The mutated gene persists in the germinal cells of this aniridic man who will eventually produce half of the germ cells with this mutant gene (Fig. 10.3).

Thus, mutation is a change in gene, potentially capable of being transmitted in the changed form. Mutation occurs in genes (gene mutation), as well as, in chromosomes (chromosomal mutation).

Mutation, therefore, is the raw material of evolution. Evolution absolutely depends on mutations because this is the only way that new 'alleles' (i.e., one of the two or more alternative forms of a gene that arise by mutation), and 'new regulatory regions' are created. But, this seems contradictory because most mutations are either harmful or neutral. However, the exact usage of mutation in evolution is debatable.

It may be summarized that evolution is driven by mutations and shaped by natural selection. Some mutations are detrimental, some are advantageous, while majority of them have no effect at all.

POINTS TO REMEMBER

- Darwinism is a theory of biological evolution proposed by Charles Darwin.

- Organic evolution is the change in the inherited characteristics of biological populations over successive generations.

- Evolutionary processes give rise to diversity at every level of living forms.

- A mutation occurs when a DNA gene is changed in such a way as to alter the genetic message carried by that gene.

- Lamarck stated that all living organisms had originated from their ancestors and while doing so they had slowly adapted for living in the available environments.

- The micro-evolution takes place within species with minor transitions, as a result of changing interactions between the animals and the environment.

- The macro-evolution, refers to evolution on higher levels.

- Lamarck states that an organism can pass on characteristics that it acquired during its lifetime to its offspring.

EXERCISE

1. Define the term 'Organic Evolution' and explain various theories related to it.
2. Define the term 'evolution' and explain its process
3. Write notes on :
   a) Macro- and Micro-evolution
   b) Evidence of evolution
   c) Darwinism
   d) Lamarckism
   e) Mutation

❖ ❖ ❖

Chapter 11...

INTRODUCTION TO PETROLOGY

Contents ...

11.1 INTRODUCTION

Definition of petrology: The large variety of rocks on the earth's surface, as well as beneath it, is made up of different minerals, mineraloids or glass. The study of rocks is called as Petrology (in Greek, *petros* means rocks; *logos* means to study).

Definition of Lithology : Lithology may be defined as the study of physical characteristics of a rock, including colour, composition, and texture.

Most rocks are made up of two or more minerals (polymineralic) but a few are made up of only one kind of mineral (monomineralic). Some rocks may also contain the fossil remains of plants and animals. The study of petrology is carried out under two major heads as follows, viz

a) **Petrography**: which describes and systematically classifies rocks, and

b) **Petrogenesis:** which studies the origin or the genesis of the rocks.

Petrology, therefore, includes the mineralogical, textural and chemical description of rocks along with their geological origin, mode of occurrence and their relation to the physico-chemical environment of the earth.

As the earth cooled from its original molten state, it gave rise to a distinct three-layered structure of the earth, having a crust, mantle and a core. The original molten rock matter is known as magma. Later, as the magma cooled, it solidified into a crystalline rock or glass or a combination of both. Geological studies show that magma has either extruded onto the earth's surface or it has cooled and solidified within the earth's interior. These rocks were the first formed rocks of the earth and hence, the igneous rocks are also referred to as the Primary rocks.

The formation of the earth's solid crust was followed by the formation of the earth's atmosphere and hydrosphere. The physical and chemical action of the atmospheric gases and water on the crustal rocks, led to their weathering, causing the rocks to break down into loose debris and also to form solutions. These products of weathering were transported by water, wind or glaciers and   deposited in the basins on land such as the lakes and rivers or into the sea.

The loose material after deposition got compacted by the weight of the overlying sediments and was cemented by the chemicals precipitated from the solutions. In this way, a large variety of rocks were derived from the pre-existing rocks and they are called as the secondary rocks.

The constant movement of the earth's tectonic plates (100 kms thick) brings about the collision or the breaking away of these tectonic plates in the earth's crust. This causes a drastic change in the temperature and pressure conditions within the rocks of the earth's crust. The rocks re-adjust to these changed physico-chemical conditions, which are manifested in the change of mineralogy and texture of the rock, giving rise to a partial or completely new rock type viz. the metamorphic rocks (In Greek, *meta* - other, *morph* - form).

Major divisions and diagnostic characteristic of rocks:

Thus, the earth's solid crust is made up of three types of rocks viz. the igneous, secondary and metamorphic with each rock type characterised by its texture, structure, mineral composition and mode of formation.

11.2 DIAGNOSTIC CHARACTERISTICS OF IGNEOUS ROCKS

(1) These are the first formed rocks of the earth's crust and hence are also called as primary rocks.

(2) Formed by the cooling of magma or lava.

(3) Usually massive and unstratified (except in lava flows).

(4) Non-fossiliferous.

(5) Often occur as veins, fissures and stringers intruding into the country rocks (host rocks).

(6) Show baking or alteration effects along their contacts with the country rocks.

(7) Show an interlocking texture consisting of minerals (crystalline) or glass (non-crystalline rock) or both, minerals and glass.

(8) The lava flows may show gas cavities and amygdales (almond shaped secondary fillings).

(9) Made up of minerals formed from very high temperature magma. Hence, these minerals are termed as pyrogenetic minerals (*pyro* - fire, *genesis* - origin).

(10) They show granitic, porphyritic and glassy textures to mention a few.

11.3 DIAGNOSTIC CHARACTERISTICS OF SEDIMENTARY ROCKS

(1) They are derived from the mechanical and / or chemical breaking down of pre-existing rocks.

(2) Most sedimentary rocks are layered and have originated by the process of sedimentation or settling of transported particles. Thus, the layering or   stratification is the single most important character of sedimentary rocks. (However, there are some igneous and metamorphic rocks which may also show stratification or pseudo-stratification).

(3) Sedimentary rocks display a clastic texture, in which individual grains are cemented together by calcium carbonate, silica or the oxides of iron. Non-clastic texture results due to chemical precipitation or biogenic processes.

(4) Sedimentary rocks under suitable conditions may contain fossils.

(5) Their mineral composition is complex as compared to the igneous and metamorphic rocks. This is because they are derived from a pre-existing rock which may have been igneous, secondary or metamorphic.

11.4 DIAGNOSTIC CHARACTERISTICS OF METAMORPHIC ROCKS

(1) Metamorphic rocks are derived by the transformation of pre-existing (igneous, sedimentary or metamorphic) rocks.

(2) They are formed in response to pronounced changes in the temperature, pressure and chemical environment of the pre-existing rocks.

(3) Mineral formation takes place essentially in the solid state, (i.e., their bulk-phase is solid).

(4) Minerals of metamorphic origin show varying degrees of recrystallisation.

(5) They may show an interlocking arrangement of crystalline material like igneous rocks.

(6) They exhibit layering, foliation or banding as a result of stress or directed pressure, which may resemble the stratification in sedimentary rocks.

(7) Most of them contain the typical metamorphic minerals such as staurolite, kyanite, sillimanite and garnet.

(8) They are generally devoid of fossils.

(9) They may show relict textures or minerals of the pre-existing rocks which were not completely transformed.

(10) Their mineral composition is simpler than that of the secondary rocks.

(11) Metamorphic structures are described as slaty, granulose, schistose and gneissose.

11.5 THE ROCK CYCLE

The weathering and erosion of the igneous rocks produces loose material which may remain in the place of its origin or may be transported, deposited and later consolidated by compaction or cementation into the sedimentary rocks. These rocks may themselves undergo weathering and erosion forming new sedimentary rocks.

The tectonic movements of the earth's lithosphere cause tremendous changes in the pressure and the temperature conditions in the igneous, secondary and metamorphic rocks, which are caught up in the plate movements, resulting in the transformation of these rocks into metamorphic rocks. When the tectonic plates undergo subduction and descend to great depths, the rocks of these plates are likely to melt partially or completely, thus, generating a magma which may later cool down producing igneous rocks. Thus, in nature, there appears to be a cyclic production, distortion and re-production of all the three types of rocks. This appears to be a continuing process ever since the earth came into existence 4.5 billion years ago. This phenomenon is described as the Rock Cycle and is diagrammatically shown in Fig. 11.1.

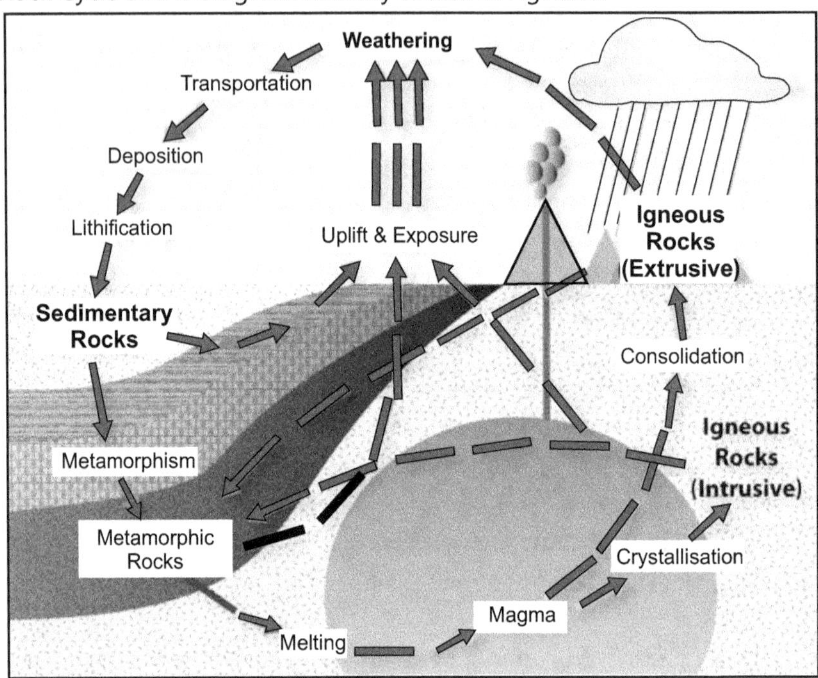

Fig. 11.1: Rock cycle

POINTS TO REMEMBER

- The study of rocks is called as Petrology which includes the mineralogical, textural and chemical description of rocks along with their geological origin, mode of occurrence and their relation to the physico-chemical environment of the earth.

- When rocks are made up of two or more minerals they are called polymineralic.

- When rocks are made up of only one kind of mineral they are called monomineralic.

- The earth's molten state, gave rise to a distinct three-layered structure, having a crust, mantle and a core.

- The earth's solid crust is made up of three types of rocks - the igneous, secondary and metamorphic rock.
- The igneous rocks are the first formed rocks of the earth's crust.
- They are formed by the cooling of magma or lava and are usually massive and unstratified.
- The secondary rocks are derived from the mechanical and / or chemical breaking down of pre-existing rocks.
- Sedimentary rocks display a clastic texture.
- They are formed in response to changes in the temperature, pressure and chemical environment of the pre-existing rocks.
- Metamorphic rocks are derived by the transformation of pre-existing (igneous, sedimentary or metamorphic) rocks.

EXERCISE

1. With the help of neat diagram explain the Rock cycle.
2. Define the term 'petrology' and give the important characteristic properties of the igneous rocks.
3. Define the term 'petrology' and give the important characteristic properties of the sedimentary rocks.
4. Define the term 'petrology' and give the important characteristic properties of the metamorphic rocks.
5. Define the terms 'primary' and 'secondary' minerals and compare the igneous and the sedimentary rocks.

Chapter **12**...

MAGMA AND ITS COMPOSITION

Contents ...

12.1 MAGMA AND ITS COMPOSITION

Observation at volcanic vents during an eruption, shows that an igneous rock is formed by the solidification of a mobile, hot and molten rock matter (produced inside the earth) giving rise to a rock such as basalt. This molten rock matter is called as magma and it is essentially a hot, thick, viscous, naturally occurring material that is generated within a planet or any other cosmic body of similar nature. **Turner** and **Verhoogen** (1962) have defined magma as a naturally occurring molten rock matter that consists of the liquid phase and having a composition of a silicate melt. Magma is considered to be a multicomponent system, consisting of a liquid phase (mutual solutions of all the minerals) and of a number of solid phases such as the suspended crystals of the early formed minerals like olivine, pyroxene and plagioclase. All igneous rocks which have been formed below the earth's surface have also cooled and consolidated from magma. The magma, when it reaches the earth's surface, gives off its gases and volatiles (due to the reduction in pressure) and then it is called lava. Thus, magma also contains a gaseous phase consisting of water (steam) with minor amounts of carbon dioxide, hydrochloric acid, hydrogen fluoride and sulphur dioxide. The volatiles occur in dissolved state under high pressures but when the magma migrates upwards and intrudes into the shallow parts of the earth's crust, they are transformed into the gaseous state. This transformation also increases the volume of the magma. As the magma reaches near the earth's surface, the increase in volume may surpass the elastic limit of the rocks into which it intrudes causing it to split apart and throws out the lava as a volcanic eruption.

The cooling and consolidation of the lava on the earth's surface gives rise to igneous rocks. Thus, lavas which are the surface representatives of magma provide a good insight into its composition. A large number of the igneous rocks have been chemically analysed, the results of which indicate that 99% of their composition includes nine major oxides viz, oxygen, silicon, aluminium, iron, calcium, sodium, potassium, magnesium and titanium; while the remaining 1% consists of the rarer elements and especially the volatiles like hydrogen, fluorine, chlorine and sulphur. The composition of an average igneous rock is represented in Table 12.1.

Table 12.1: Major elements and their oxides in an igneous rock

Element	Percentage	Oxides	Percentage
Oxygen	46.59	SiO_2	59.12
Silicon	27.72	Al_2O_3	15.34
Aluminium	0.813	Fe_2O_3	03.08
Iron	05.01	FeO	03.80
Calcium	03.63	MgO	03.49
Sodium	02.85	CaO	05.08
Potassium	02.60	Na_2O	03.84
Magnesium	02.09	K_2O	03.13
Titanium	00.63	H_2O	01.15
Phosphorus	00.13	CO_2	01.10
Hydrogen	00.13	TiO_2	01.05
Manganese	00.10	ZrO_2	0.04
Sulphur	0.052	P_2O_5	0.30
Barium	0.050	Cl	0.05
Chlorine	0.048	F	0.03
Chromium	0.032	S	0.05
Carbon	0.032	Fluorine	0.030

The analysis of an igneous rock does not necessarily always represent the original composition of the magma especially as the volatiles tend to escape from the cooling magma. But observations at recent volcanic eruptions and the collection of gases from the lavas have clearly indicated that a good amount of volatiles are present in the magma as compared to the volatiles recorded in the igneous rock analysis. Water alone is found to make up 4% of the basaltic magmas.

12.2 PYROGENETIC MINERALS

Minerals formed by the cooling of magmas or lavas are called as pyrogenetic minerals (i.e., formed from fire). Oxygen and silicon are the most abundant elements in the magmas and therefore, silicates and silica constitute the major bulk of the pyrogenetic minerals. Other oxides occur in much lesser amounts but they are important from a petrogenetic point of view.

The formation of the silicate minerals in magma is due to the three types of acids which are formed by silicon, viz.

(1) Orthosilicic acid - H_4SiO_4 or $(2H_2O, SiO_2)$

  e.g., Olivine 2 $(Mg, Fe)O, SiO_2$.

(2) Metasilicic acid - $H_4Si_2O_6$ or $(2H_2O, 2SiO_2)$

  e.g., Enstatite $(Mg, Fe)OSiO_2$.

(3) Polysilicic acid - $H_4Si_3O_8$ or $(2H_2O, 3SiO_2)$

  e.g., Orthoclase $K_2OAl_2O_36SiO_2$.

The pyrogenetic silicates, therefore, consist of three groups viz., the orthosilicates, the metasilicates and polysilicates having an oxygen ratio of bases to silica as $1 : 1$; $1 : 2$ and $1 : 3$ respectively. Minerals like olivine are orthosilicates; pyroxenes (e.g., augite) and amphiboles (e.g., hornblende) are metasilicates while the feldspar group minerals (e.g., orthoclase and albite) are polysilicates.

The chemical composition of these silicate minerals reveals that potassium and sodium form the most active bases present in a magma, followed by the less active calcium and then, by the least active magnesium and iron. The effect of this varied relative activity of the bases is that silica is taken up to its fullest extent by potassium and sodium to form polysilicates (alongwith aluminium), calcium tends to form metasilicates, whereas magnesium and iron give rise to orthosilicates.

Iron, which has the least affinity for silica, very often appears as an iron oxide e.g., the iron mineral magnetite.

If magma is rich in silica it forms the polysilicate and metasilicate minerals. If the magma is deficient in silica then the alkalis, instead of forming the feldspars (e.g., orthoclase and albite), form the feldspathoid minerals, as these minerals require less silica for their formation. On the same lines, magnesium and iron give rise to olivine rather than the silica rich pyroxenes.

A magma having significant amount of water (wet magma), will give rise to minerals which contain the hydroxyl molecules such as the amphiboles and the micas. A relatively 'dry' magma forms the pyroxenes and olivines.

The excess silica which may be present in the magma, after having fully satisfied all the bases, crystallises out as quartz.

Based on the silica content of a magma the pyrogenetic minerals can be classified into two types

(1) The minerals of low silication and

(2) The minerals of high silication

Minerals of low silication: Leucite, nepheline, olivine and biotite.

Minerals of high silication: Orthoclase, albite, augite and hornblende.

The mineral quartz generally does not occur in a rock which contains minerals of low silication. The minor chemical constituents of magma give rise to minerals of very small size. These minerals usually possess well-developed crystalline forms and also have wide distribution in the rocks. These minerals are appropriately called as the accessory minerals and common examples are ilmenite $(Fe,Ti)_2O_3$; apatite $Ca(PO_4)_2$; sphene $CaOTiO_2$ and zircon (ZrO_2SiO_2). Some metamorphic minerals like kyanite, sillimanite, cordierite and staurolite never occur as pyrogenetic minerals.

12.3 BOWEN'S REACTION SERIES

Bowen's reaction series basically explains how igneous rocks form while cooling of magma takes place. Norman Bowen, a Canadian geologist, through his laboratory experiments and field observations, discovered that minerals crystallize differently from the magma during its cooling process, as a crystallization sequence. As magma cools, specific minerals are formed at specific temperatures. He also predicted that the reaction relation between the minerals that crystallise and the surrounding magma is of two types. As magma cools and if the minerals crystallise as a result of simultaneous crystallisation without reaction between the minerals (eutectic crystallization), then these minerals do not react further with the magma around them. On the other hand, some minerals are constantly reacting with the magma around them and their compositions are continually getting modified. Hence, he proposed two sequences namely, 'discontinuous' and 'continuous' reaction series.

Discontinuous Reaction Series:

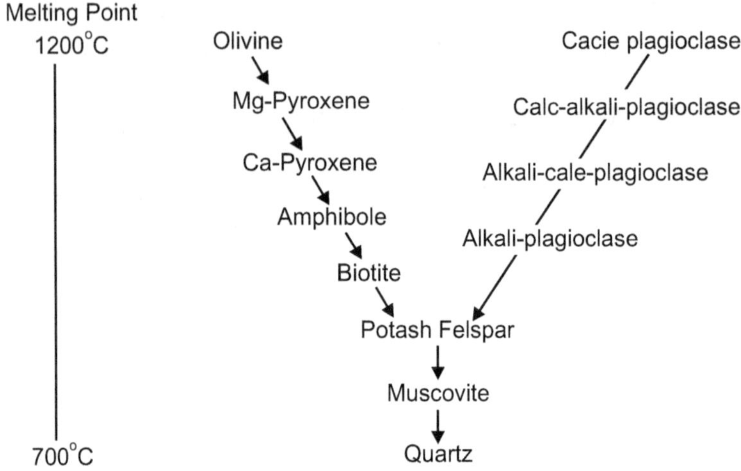

Fig. 12.1

The discontinuous branch describes the formation of minerals like olivine, pyroxene, amphibole, and biotite which are more enriched in iron and magnesium. As cooling takes place, at very high temperatures, olivine is the first mineral to crystallise. Then, as magma begins to cool, pyroxene crystallises. With more cooling, amphibole crystallises and finally biotite appears within the magma.

The discontinuous series progresses with a distinct change in the composition of minerals (i.e., creation of new minerals). As this change is not a smooth continuous flow, it is termed as a discontinuous process. After the formation of biotite, the remaining residual magma continues to cool and forms potassium feldspar, muscovite or quartz.

Continuous Reaction Series:

The continuous reaction series is also developed within the magma at the same time as the discontinuous series (Fig. 12.1). In contrast to the discontinuous series, the minerals of this series have a flow or 'continual' reaction taking place, and hence the name. Within the continuous series, plagioclase minerals crystallise with varying compositions (i.e., mix-crystals). It starts with the highest temperature mineral, which is calcium-rich plagioclase.

As the magma cools down, the calcium within the plagioclase minerals is replaced with sodium. Hence, this gradual replacement is from the calcium-rich to sodium-rich plagioclases. Therefore, a plagioclase with 50% calcium and 50% sodium develops when the continuous series reaches its middle stage. Finally, at the end of this series, sodium-rich plagioclase crystallises out. Further, as magma cools and the chemical characteristics continue to change, there is a formation of potassium feldspar, muscovite or quartz.

In this reaction series, the mineral placed 'below' has been produced after the reaction between the mineral 'above' it and the magma. Both the branches of the reaction series finally converge into a single discontinuous series, i.e., potash feldspar, muscovite and quartz.

12.4 FORMATION OF CRYSTALS AND GLASS

An igneous rock consists of crystalline and / or non-crystalline matter. The crystalline matter consists of minerals which have a definite internal atomic structure, and a well ordered arrangement of its molecules with a dense packing. The formation of crystals is favoured by a slow rate of cooling of the magma under high pressure.

On the other hand, a non-crystalline matter or glass represents a rapidly cooled liquid which has formed an amorphous solid. Its molecules occur at random with a disordered orientation. This type of a solid having a low density is called as a geological glass. The formation of glass, therefore, is favoured by a fast rate of cooling of lava at low pressure.

Thus, it is clear that the slow cooling of magma at great depths produces crystalline matter whereas the formation of glass takes place at or near the earth's surface.

POINTS TO REMEMBER

- The hot and molten rock matter produced inside the earth is called as magma.
- Magma is a multicomponent system, consisting of a liquid phase and of a number of solid phases such as the suspended crystals of the early formed minerals.

- Lava is the surface representatives of magma and provides a good insight into its composition.

- Minerals formed by the cooling of magma or lava are called as pyrogenetic minerals.

- Silicates and silica constitute the major bulk of the pyrogenetic minerals.

- The pyrogenetic silicates consist of three groups - the orthosilicates, the metasilicates and polysilicates having an oxygen ratio of bases to silica as 1:1; 1:2 and 1:3 respectively.

- Based on the silica content of a magma the pyrogenetic minerals are classified into- The minerals of low silication and the minerals of high silication.

- The formation of crystals is favoured by a slow rate of cooling of the magma under high pressure.

- The formation of glass is favoured by a fast rate of cooling of lava at low pressure.

EXERCISE

1. Write notes on:
    - (a)　Magma and lava
    - (b)　Minerals of high and low silication
    - (c)　Formation of crystals and glass
    - (d)　Pyrogenetic minerals.

Chapter **13**...

FORMS OF IGNEOUS BODIES

Contents ...

13.1 INTRODUCTION

The forms of igneous rocks are broadly classified into two types based on their mode of occurrence, size, shape and relation with the country rock.

They are:

1) **Intrusive:** i.e., rocks formed by the subsurface emplacement of magma

2) **Extrusive:** i.e., rocks formed on the earth's surface

The shape, size, texture and structures of the rocks and their relation with the host rocks, help to distinguish the various forms shown by the intrusive and the extrusive igneous rocks.

Intrusive rocks rarely show the presence of glass and vesicles as they crystallise under a slow rate of cooling. This gives rise to coarse mineral grains as compared to the lavas. Intrusive igneous rocks, further, are found to affect and alter the country rocks on all sides where they are in contact whereas lavas exhibit baking effects only on the top of the rocks on which they flow.

Extrusive rocks are developed either due to the cooling and solidification of lava or they could be the accumulation of pyroclastic matter (solid products) ejected during volcanism. The lava loses most of its volatile and gaseous content when extruded on the earth's surface and the resulting rock shows a surface, characteristically cindery, cloggy and vesicular. The presence of glass in the extrusive rocks is indicative of their rapid cooling while flow structures suggest their flowage before solidification.

13.2 INTRUSIVE ROCKS

They are formed by the intrusion or the injection of magma into the host or country rocks. The magma is forced through cracks, joints or the bedding planes of the host rocks. The shape, size and the relation of the intrusive with the structure of the host rock, is used to describe the forms of the intrusive rocks.

The intrusive igneous rock is coarse to medium grained and does not have vesicles, nor does it show flow structure. The contacts of the intrusive with the host rocks show baking effects.

The forms of intrusive igneous rocks are sub-divided into two types, viz.

(1) Concordant: i.e., there is a rough parallelism between the intrusive and the major structure of the host rock (like bedding planes).

(2) Discordant: Where the intrusive cuts across the major structure (e.g., bedding planes) of the host rock. These have been shown in Fig. 13.1 and their tabular classification is given in Table 13.1.

Table 13.1: Intrusive rocks

Concordant	Discordant
e.g., sill, laccolith and lopolith	e.g., dyke, veins and batholith

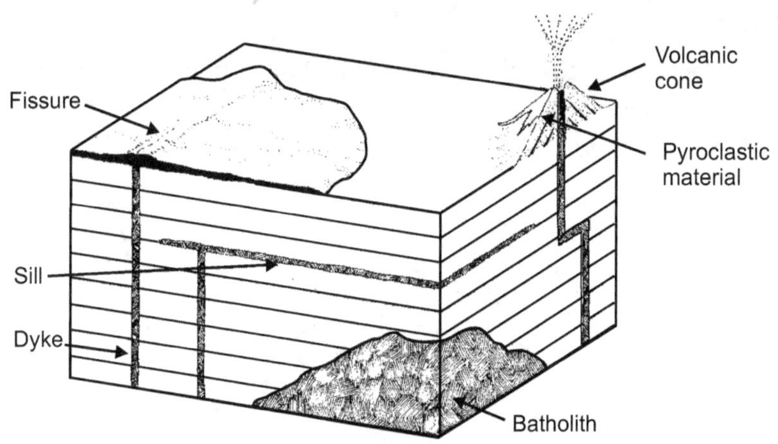

Fig. 13.1: Structure of a sill, batholith, dyke and stock

13.2.1 Concordant Intrusions

Sill:

Sills are sheet-like concordant injections of magma, which have intruded along the bedding planes or other planar surfaces of the country rock (Fig. 13.1).

A sill may have a width of a few centimeters to several meters, with its thickness being negligible as compared to its lateral extent. Sills are reported in the basic rocks like basalt. They generally occur at shallow depths along the bedding planes of stratified rocks. They are formed when the pressure of the intruding magma is slightly greater than the pressure exerted by the weight of the overlying rock. The magma spreads laterally along the low pressure areas like the bedding planes. The upper and lower boundaries of the sill are more or less parallel for a considerable length, before they finally pinch out at their lateral extremities.

Sills are classified as 1) simple or 2) composite, depending upon their chemical composition. Simple sills are homogeneous intrusive bodies while the composite or multiple sills are heterogeneous in their chemical composition.

Laccolith:

A laccolith is a mushroom- or bun-shaped body which is formed when a highly viscous magma intrudes along a plane of weakness in a rock forming an arched roof and a flat base or floor (Fig. 13.2).

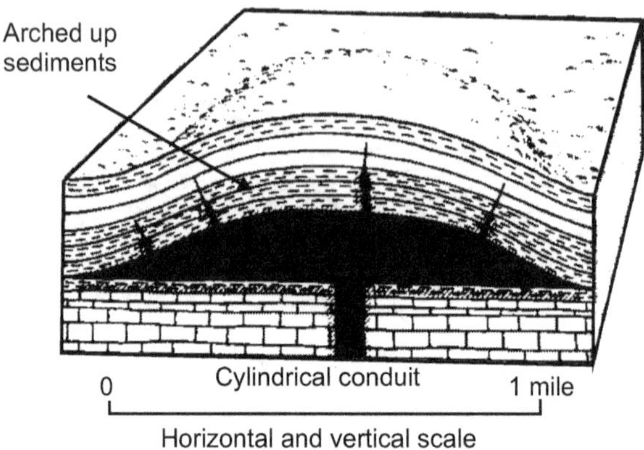

Fig. 13.2: A cross-section of a laccolith

A laccolith has a circular or elliptical shape with a small lateral extent as compared to its vertical thickness. The laccoliths are connected to the magma chamber by a central pipe and feeder channels.

Lopolith

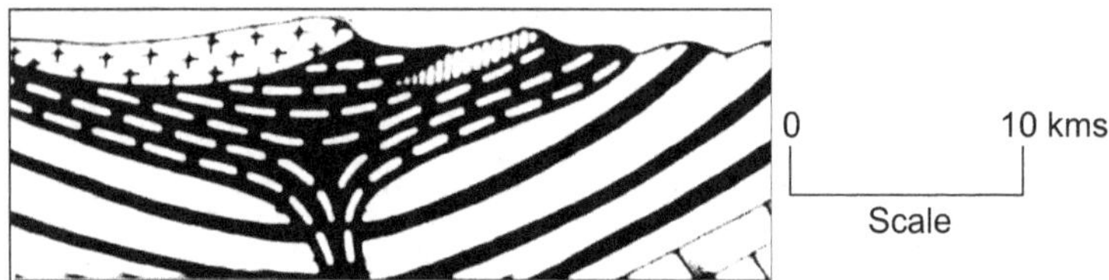

Fig. 13.3: A Lopolith

In Greek, *Lopas* means shallow basins. Lopoliths are usually thick, saucer shaped igneous intrusions which have sunken centers and elevated rims (Fig. 13.3).

Lopoliths have a geometry which is opposite to that of a laccolith and they are also of a much larger dimension as compared to the laccoliths. Lopoliths are generally layered and the rock is basic in composition. The layering results from a different mineralogy and / or texture. Usually the thickness of lopoliths is one-tenth of their diameter, e.g. Lopoliths of the Bushveld complex of South Africa have an area of 6,600 sq. kms.

13.2.2 Discordant Intrusions

Dykes and Veins:

Dykes are intrusions which are vertical or near vertical at the time of their intrusion. Although they did not reach the earth's surface during their emplacement, today they may be exposed on the ground after removal of the overlying rocks due to erosion. The thickness of dykes varies from a fraction of a centimeter to several tens of meters whereas their length may vary from a few meters to hundreds of meters. Dykes usually intrude along fractures or joint systems of the country rocks, e.g., dolerite dykes in the basalt rocks of Murud Janjira (Maharashtra). Because dykes are elongated in one direction, they are easily identified as straight or curvilinear lines on aerial photographs. Also, as a result of differential weathering, dykes form either topographic elevations, flat topography or depressions on the earth's surface (Fig. 13.4).

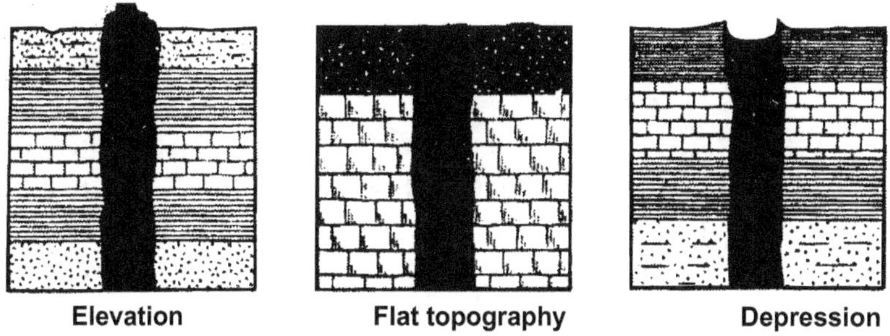

Fig. 13.4: Topographic modifications due to dykes

Along shear zones or the weaker zones of the earth's surface, dykes may occur as a cluster described as a swarm of dykes, e.g., east-west trending dyke swarms of basic composition along the river Narmada.

Multiple dykes are formed when successive intrusions of the same or varying compositions are forced through the same dyke fissure, while a composite dyke develops when magma of varied composition (i.e., acidic and basic) intrudes along the same fissure.

During their emplacement, dykes develop chilled margins along their contacts with the country rock. These margins are fine grained or glassy while the central portion of the dyke is coarse grained.

Ring Dykes:

When magma reaches near the earth's surface and is intruded into the country rocks it has left behind a void, into which the overlying rocks may collapse or sag. This causes annular fractures to develop, which may later get filled up with magma, giving an arcuate system of intrusions described as ring dykes (Fig. 13.5).

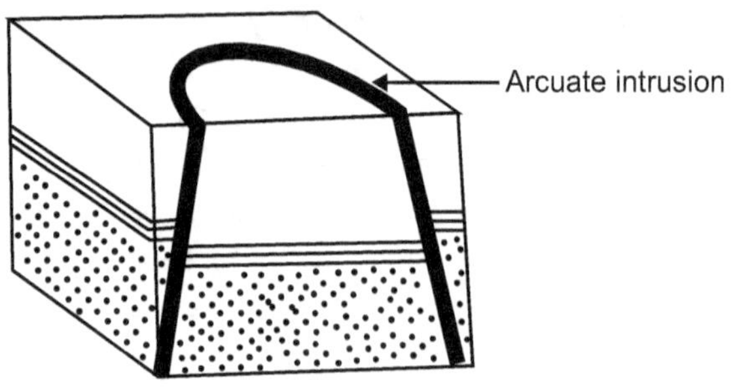

Fig. 13.5: Ring dyke

Veins

It is a distinct but thin sheet-like body of crystallised minerals within a rock. Veins form when mineral constituents carried by magmatic solutions within the rock mass are deposited through precipitation. This occurs when a magmatic emanations reach the overlying rocks through small vertical or near-vertical channels (Fig. 13.6). These are usually composed of silica and may occur in the form of closely spaced configuration.

Fig. 13.6: Veins in the rocks

Batholiths:

A batholith(In Greek, *bathos* means deep) is a deep seated intrusion which has steep walls and irregular shapes (Fig. 13.1). Batholiths originate at plutonic depths and hence their floors are not traceable and they are exposed on the surface only after very intense erosion of the overlying rocks over geological time. They cover very large areas measuring upto several hundred kilometers. Batholiths occur in orogenic belts, especially along the tectonic axes of the mountain chains and they are generally granitic in composition, e.g., granitic batholiths of Mount Abu, Rajasthan. The term pluton is preferred to that of 'batholith' by many geologists due to its uncertain genesis and variable geometry. A smaller batholith is often referred to as a stock, while the term boss is used for a circular outcrop of a stock. Stocks and bosses are probably the appophyses of the main batholith.

13.3 FORMS OF EXTRUSIVE IGNEOUS ROCKS

The magma when extruded on the surface loses its gases and volatiles due to the drop in pressure and it is then called lava. In contact with the atmosphere, the lava cools rapidly, producing a rock which is fire grained, with a likely presence of glass and which has gas cavities, known as vesicles, at the base of the flow. The surface of the flow may appear cindery and rough as a result of the sluggish flow of a viscous magma. A smooth, ropy surface results from a fast moving and mobile magma.

Usually the hot lava develops a 'baked' contact with the rock surface over which it flows.

Each of these characters is typical of extrusive rocks and the presence of any of them helps to differentiate it from intrusive rocks.

There are two types of channels along which a lava may be extruded - 1) Central type of eruption, where the lava issues out of a pipe or conduit. 2) Fissure type of eruption, where large volume of lava comes out of long fissures or cracks.

Central type of eruption: The lava is erupted from a pipe and it accumulates around the mouth or the orifice of the pipe. Repeated eruptions result in the building up of a cone and crater structure of a volcano (Fig. 13.7).

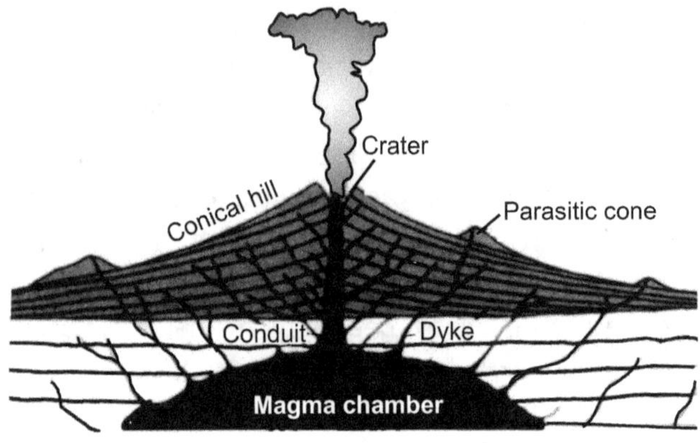

Fig. 13.7: Cone and crater and parasitic cone

The area affected by such a volcanic eruption is generally small. The lava may contain pyroclastic material and ash. arcuate system of intrusions, which are acidic in nature, are very viscous and hence, their movement is sluggish. They tend to consolidate in steep sided bulbous masses around the eruptive pipe, known as a *Puy*, e.g., volcanoes observed in the Pacific ocean, in Europe and Andamans.

Fissure type of eruption: This type of eruption takes place through long cracks in the earth called fissures, from which voluminous amount of lava is poured out on both sides of the fissure. Lavas of basic composition (containing iron and magnesium minerals in abundance) are very mobile and hence, spread over considerable distance (Fig. 13.8). The enormously thick basalts of western India have been formed by fissure eruptions during the Cretaceous - Eocene period. They are spread over an area of about 50,000 sq. kms.

Fig. 13.8: Fissure type of volcanic eruption

POINTS TO REMEMBER

- The igneous rocks are broadly classified into intrusive and extrusive rocks.
- Intrusive rocks are formed by the subsurface emplacement of magma.
- Extrusive rocks are formed on the earth's surface.
- Intrusive rocks are formed by the intrusion or the injection of magma into the host or country rocks.
- The intrusive igneous rock is coarse to medium grained. They do not have vesicles, nor do they show flow structure.
- The forms of intrusive igneous rocks are sub-divided into two types - concordant and discordant.
- In concordant there is a rough parallelism between the intrusive and the major structure of the host rock.
- In discordant the intrusive rocks cuts across the major structure of the host rock.
- Sills are sheet-like concordant injections of magma, which have intruded along the bedding planes or other planar surfaces of the country rock
- Sills are classified as simple or composite.

- Simple sills are homogeneous intrusive bodies while the composite or multiple sills are heterogeneous in their chemical composition.
- A laccolith is a mushroom- or bun-shaped body which is formed when a highly viscous magma intrudes along a plane of weakness in a rock forming an arched roof and a flat base or floor.
- Lopoliths are generally layered and the rock is basic in composition.
- Dykes are intrusions which are vertical or near vertical at the time of their intrusion.
- Veins form when mineral constituents carried by magmatic solutions within the rock mass are deposited through precipitation.
- A batholith is a deep seated intrusion which has steep walls and irregular shapes.
- The magma when extruded on the surface loses its gases and volatiles due to the drop in pressure and it is then called lava.
- There are two types of channels along which a lava may be extruded - Central type of eruption, where the lava issues out of a pipe or conduit. And Fissure type of eruption, where large volume of lava comes out of long fissures or cracks.

EXERCISE

1. Write notes on

    (a) Forms of extrusive igneous rocks

    (b) Forms of intrusive igneous rocks

    (c) Lopoliths and laccoliths

    (d) Volcanic plug and ring dykes

Chapter **14**...

TEXTURES & STRUCTURES
OF IGNEOUS ROCKS

Contents ...

14.1 INTRODUCTION

Texture describes the mutual relation of the rock constituents (minerals and glass) within a uniform aggregate. It takes into account the relative amount of crystalline and glassy matter as well as the size, shape and the arrangement of the minerals. Thus, the **texture** of a rock is defined as the 'intimate mutual relations of the minerals and glassy matter in a rock made up of a uniform aggregate'.

Rock textures are significant characters of a rock as they throw light on the genesis of the rock and the environmental conditions under which it formed. The **structure** of a rock represents the juxta-position of two or more textural aggregates in a rock. Usually, structures represent large scale features like pillow lavas. But, some structures studied under the microscope are described as micro-structures. Textures and micro-structures are best studied under the microscope and they help in unravelling the geological processes and the physico-chemical conditions under which the rocks formed.

14.2 FACTORS CONTROLLING TEXTURES OF IGNEOUS ROCK

The four inherent factors which control the texture of a rock are :

(1)   Crystallinity

(2)   Granularity

(3)   Shape of the crystals and

(4)   The mutual relationship of the crystals or between the crystals and glass

The factors 3 and 4 are grouped together as the 'fabric' of the rock. Thus, the texture of an igneous rock is said to be a function of its crystallinity, granularity and fabric.

1. Crystallinity: This property expresses the proportion of the crystalline and the glassy material in an igneous rock. Some igneous rocks such as granites are made up entirely of minerals and are called holocrystalline; other rocks like obsidian which contain only glass are known as holohyaline. Many rocks such as basalt may contain both minerals and glass and are called as *hypo-* or *mero*-crystalline (Fig. 14.1).

Crystallinity of a rock depends upon

(i)   the rate of cooling of the magma and

(ii)   the viscosity of the magma from which it formed.

A slow rate of cooling and a low viscosity of the magma, favours the formation of crystals and such conditions exist at great (plutonic) depths in the earth. Hence, a holocrystalline rock like granite is formed at plutonic depths. A fast cooling rate and a high viscosity of magma / lava exist at shallow (intermediate) depths or on the earth's surface, which give rise to the formation of glass. Thus, a holohyaline rock forms dykes or sills at intermediate depths or lavas on the earth's surface.

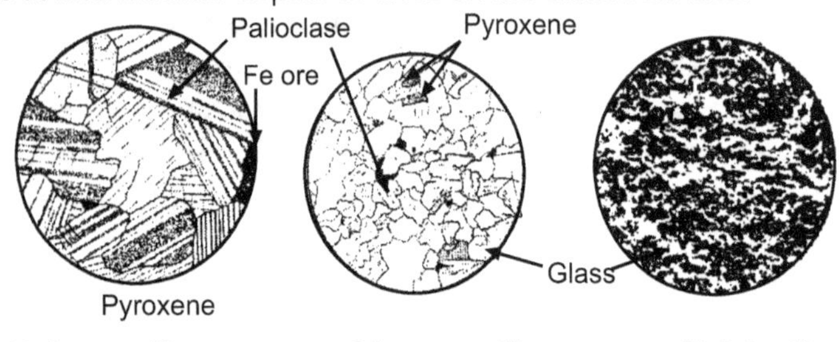

| Holocrystalline | Merocrystalline | Holohyaline |

Fig. 14.1: Degree of crystallinity in igneous rocks

2.   Granularity: The grain size of igneous rocks varies widely and is dependant primarily on the rate of cooling of the magma or lava. Rocks are described as phaneric or phanerocrystalline if their mineral grains can be distinguished individually with the naked eye. This may not be possible in the fine grained rocks in which the mineral grains cannot be distinguished by the naked eye and such rocks are described as aphanitic.

Phaneric rocks are further divided on the basis of their average mineral grain diameter into fine grained (with grain diameter less than 1 mm), medium grained (between 1-5 mm), coarse grained (between 5-30 mm) and very coarse grained or pegmatitic (grain diameter more than 3 cms).

Aphanitic rocks are described as microcrystalline when the individual grains can be distinguished only under a microscope and as cryptocrystalline, when the individual grains are indistinguishable even under the microscope.

Besides the rate of cooling and the viscosity of the magma, the granularity of the rocks also depends on the molecular concentration of the mineral constituents in the magma. A large molecular concentration gives rise to larger grains of the mineral, while a small molecular concentration gives rise to smaller grains.

3.   Shapes of the mineral grains / crystals: It is an important criterion for describing and interpreting the texture of the rock. In a holocrystalline rock, if the mineral grain shows well developed crystal faces it is termed as euhedral and a rock containing mostly euhedral grains shows an idiomorphic texture. If only some crystal faces are developed the grains are termed as subhedral and the rock containing mainly subhedral crystals is said to exhibit a hypidiomorphic texture. If no crystal faces are developed by the mineral at all (i.e., when the mineral outline is irregular), the grain is termed as anhedral and the texture exhibited by a rock containing a large number of anhedral grains is called as allotriomorphic (Fig. 14.2).

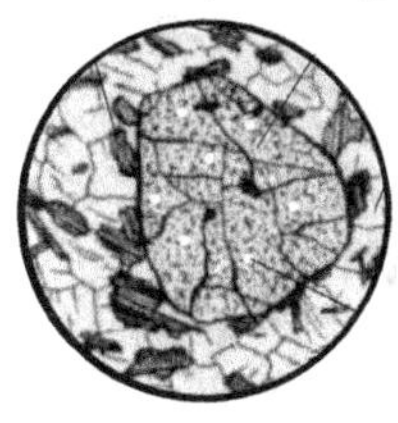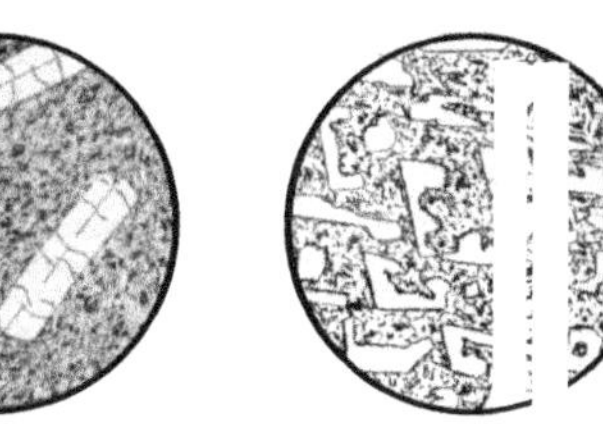

Idiomorphic   Hypidiomorphic   Allotriomorphic

Fig. 14.2: Textures on the basis of the shapes of the grains

The shape of a mineral in an igneous rock is partly controlled by its atomic structure and partly by the conditions under which growth took place. The growth of a euhedral crystal is favoured in a thinly liquid magma, which allows sufficient freedom of movement for the molecules and little interference from other mineral grains. The shape of the mineral is often decided by its position in the sequence of crystallisation of the mineral from the magma. The earliest formed minerals, therefore, form euhedral shapes while later minerals encounter interference from the many grains around it and hence are forced to assume irregular shapes as per the available space. Olivine and augite are good examples of euhedral minerals along with the early forming accessory minerals, apatite, zircon and rutile. Rapid cooling in dry magmas often yields anhedral grains, while the rapid crystallisation of viscous magmas produces radiating clusters of needle shaped grains viz. spherulites.

4.   Mutual relationship of the grains / crystals: A study of the distribution of mineral grains, their shapes and the mutual relation between these grains or between the grains and glass provides information about the fabric of the rock. Based upon the type of fabric, several textures have been recognised viz. equigranular, inequigranular, intergrowth, flowage or directive. These textures are also known as intergrain textures as they display a relation between two grains.

(a) Equigranular texture: When the major minerals in a rock are more or less of equal size, the rock is said to show an equigranular texture. An equigranular rock may be either an idiomorphic, hypidiomorphic or allotriomorphic depending upon the abundance of the euhedral, subhedral or anhedral grains, respectively (Fig. 14.3).

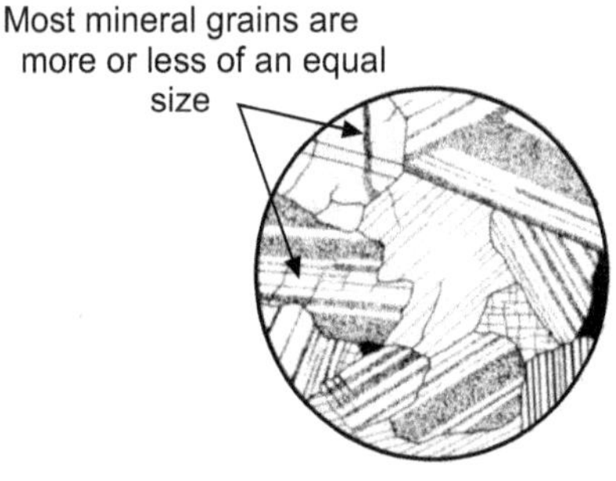

Fig. 14.3: Equigranular texture

Equigranular texture is also known as granitic texture as it is typically shown by the rock granite. This texture is generally developed in the plutonic rocks which have solidified under high pressure and a slow cooling rate. Fine grained volcanic rocks, when viewed under the microscope, may appear hypidiomorphic and are then described as micro-granitic.

(b) Inequigranular texture: When the grains in a rock show a marked difference in their grain size, the texture is said to be inequigranular. Some inequigranular textures are described as porphyritic and poikilitic, based on the fabric of the rock. In the porphyritic texture, the large grains known as phenocrysts are embedded in a groundmass which is either fine grained crystalline, cryptocrystalline or glassy (Fig.14.4). Porphyritic texture is shown by the volcanic rock basalt and the hypabyssal rock dolerite. A basalt consists of plagioclase, olivine and / or augite grains as phenocrysts and they may also form smaller grains in the groundmass. The development of a texture with different grain size can be explained as the result of two stages of crystallisation of the cooling magma from which the rock is formed.

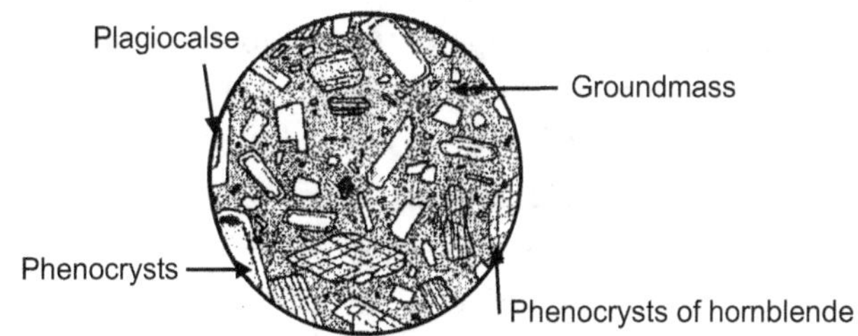

Fig. 14.4: Porphyritic textures

A slow rate of cooling during the initial crystallisation of the magma at plutonic depth gave rise to the formation of phenocrysts. When the magma carrying these phenocrysts moved upwards to shallower depths or to the earth's surface, the rate of cooling increased rapidly enabling the formation of a fine grained aggregate of minerals and / or glassy matter, which forms the groundmass in the rock. When porphyritic texture is clearly seen in hand specimens, it is described as a mega-porphyritic texture (e.g., megaporphyritic basalt).

(c) Glassy texture: When magma intrudes the country rock at very shallow depths or if it is extruded onto the earth's surface as a lava, it cools down very rapidly. This happens because of direct contact with the atmosphere, or with water or with the country rock. The rapid cooling leads to the formation of a non-crystalline solid known as glass (Fig. 14.5). Under the microscope, a glass appears isotropic but may sometimes show very fine mineral grains (e.g., plagioclase). Based upon the water content, the glassy rocks are called as pitchstone, obsidian and tachylite.

Fig. 14.5: Glassy texture in (a) pitchstone and glassy structure in (b) obsidian

14.3 STRUCTURES IN IGNEOUS ROCKS

The large scale, easily recognisable igneous textural features visible to the naked eye are known as structures. Blocky and ropy lavas, pillow, flow and joints are the major structures shown by igneous rocks. Like textures, the structures also provide valuable information on The Genesis Of The Rock. Smaller Structures Like Vesicles And Amygdales Result From The Juxtaposition Of Two Textures In A Single Rock.

(a) Vesicular and Amygdaloidal Structures:

These are common in many volcanic rocks and develop when the magma reaches the earth's surface. The low pressure at or near the surface allows the partial release and expansion of the dissolved water and / or other volatiles. Due to this, steam bubbles form which are preserved as small cavities on the rock surface after the magma consolidates. These cavities or voids may be spherical, sub-spherical or ellipsoidal and of differing sizes. The cavities are called vesicles and the rock is said to show a vesicular structure. Basalt often exhibits this structure. In highly viscous lavas (e.g., rhyolitic), much gas may be entrapped, but only tiny bubbles may form. Rapid cooling of this frothy liquid produces a pumaceous structure, characteristically shown by the rock pumice. In less viscous lavas (e.g., basaltic), integration of tiny bubbles produces a coarser, spongy or scoriaceous structure (e.g., scoria).

Vesiculation of some basaltic lavas produces well-formed ellipsoidal cavities, which may later be filled with secondary minerals like quartz, calcite or zeolite. These secondary minerals crystallise from fluids which entered the rock long after it had solidified and are called amygdales, as they are almond shaped. The rock is then said to show amygdaloidal structure. Some of the basaltic rocks preserve pipe-shaped cavities formed by the release of gases trapped in the rock over which the extruded lava flows. Later, these pipes may also get filled up with secondary minerals or ferromagnesian silicates called as 'green earths'. Due to its shape, it is called as pipe-amygdaloidal structure.

(b) Blocky Lava and Ropy Lava:

The surfaces of two different lavas may not appear similar and even the surface of a single lava may appear very different at different places. This non homogeneity may be due to variations in viscosity, chemical composition and differences in the rate of cooling. Sometimes, the surface appears very rough and irregular like angular blocks of varying sizes. It gives the effect of a tumbling, crowded mass of rock described appropriately as a blocky lava. In Hawaii, such blocky lavas are called as aa flows. Some very mobile lavas, on the other hand, solidify with smooth surfaces, which exhibit wrinkled or ropy forms similar to those seen on flowing tar (Fig.14.6). They show radial cracks and are described as ropy lavas and as 'pahoehoe' in Hawaii. Ropy structure is well exhibited by basalt flows.

Fig. 14.6: A Ropy lava

(c) Pillow Structure:

This structure is a peculiar feature of soda rich basalt lava flows known as the spilites. Rocks showing these structures appear to be made up of closely packed pillow shaped masses upto several feet across. Individual pillows have a fine grained crust which carries abundant vesicles commonly arranged concentrically on the pillow surface. Spilitic lavas when extruded on the sea floor and in direct contact with water, develop a lava front which rapidly congeals forming a thin crust. Later pulses of lava, break the thin crust at its convex end, giving rise to another bulbous mass or pillow. Several such pillows give rise to a pillow structure (Fig. 14.7).

Fig. 14.7: Pillow structure in volcanic rocks

(d) Flow Structure:

The lava flows are often characterised by streaks of dusty, opaque minerals, small embryonic-crystals (microlites) and elongated shreds of glass which are suggestive of the direction of flow of the lava.

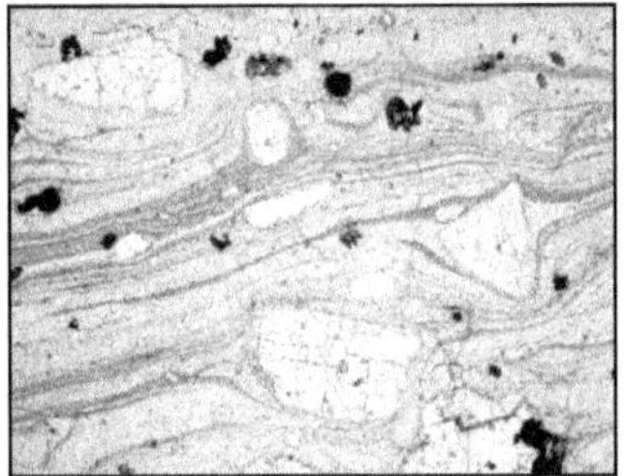

Fig. 14.8: Flow texture in rhyolite

This pattern is known as directive or flow structure when observed in hand specimen and as flow-texture, when seen under the microscope. For example, in the rock trachyte, crystals of sanidine are arranged parallel to each other in one direction, while in rhyolite the flow texture is seen in the form of parallel bands (Fig. 14.8).

(e) Columnar Joints:

Fig. 14.9: Columnar joints in a basalt flow

Joints are the fracture planes in a rock along which no movement has taken place. They are produced by the action of external forces either during crystallisation or after the rock has solidified and are seen in all kinds of igneous rocks. Joints are usually straight and smooth but sometimes they may be curved. A peculiar type of joint, known as columnar joint, is characteristic of some basaltic lava flows, dykes and sills, especially where the rock is homogeneous and fine grained. These joints are hexagonal in cross section and extend vertically from the bottom of the flow to its top enclosing polygonal areas and hence, the name 'columnar' (Fig. 14.9).

Columnar joints probably develop with the shrinkage of the fast cooling rock surface, setting up tensional stresses in the rock. Due to this stress, numerous, uniformly spaced centers of contraction develop at the corners of equilateral triangles. Shrinkage results in the development of tensional cracks, at right angles to the stress direction, giving a hexagonal pattern of vertical columns, in the case of a lava; and horizontal columns, in the case of a vertical dyke.

POINTS TO REMEMBER

- The texture of a rock is defined as the 'intimate mutual relations of the minerals and glassy matter in a rock made up of a uniform aggregate'.
- The structure of a rock represents the juxta-position of two or more textural aggregates in a rock.
- Crystallinity expresses the proportion of the crystalline and the glassy material in an igneous rock.
- Igneous rocks which are made up entirely of minerals are called holocrystalline.
- Rocks which contain only glass are known as holohyaline.
- Rocks which may contain both minerals and glass and are called as *hypo-* or *mero-*crystalline.
- Rocks are described as phaneric or phanerocrystalline if their mineral grains can be distinguished individually with the naked eye.
- Fine grained rocks in which the mineral grains cannot be distinguished by the naked eye are described as aphanitic.
- In a holocrystalline rock, if the mineral grain shows well developed crystal faces it is termed as euhedral.

- A rock containing mostly euhedral grains shows an idiomorphic texture.
- If only some crystal faces are developed the grains are termed as subhedral.
- The rock containing mainly subhedral crystals is said to exhibit a hypidiomorphic texture.
- If no crystal faces are developed by the mineral at all (i.e., when the mineral outline is irregular), the grain is termed as anhedral.
- The texture exhibited by a rock containing a large number of anhedral grains is called as allotriomorphic.
- When the major minerals in a rock are more or less of equal size, the rock is said to show an equigranular texture.
- Blocky Lava and Ropy Lava due to variations in viscosity, chemical composition and differences in the rate of cooling results in the rock surface appearing very rough and irregular like angular blocks of varying sizes.
- Pillow structure is a peculiar feature of soda rich basalt lava flows known as the spilites.
- Joints are the fracture planes in a rock along which no movement has taken place.

EXERCISE

1. Write notes on
    (a) Flow texture
    (b) Porphyritic texture
    (c) Glassy texture
    (d) Pillow and Flow structures
    (e) Vesicular and amygdaloidal structures
    (f) Granularity
    (g) Crystallinity
    (h) Equigranular and inequigranular textures
    (i) Columnar structure in igneous rocks

Chapter **15**...

CLASSIFICATION OF IGNEOUS ROCKS

Contents ...

15.1 INTRODUCTION

There are several schools of thought regarding the classification of igneous rocks. In the days before the invention of the petrological microscope and thin section studies, the rocks were classified mainly on the basis of their megascopic characters. Later, with the advancement of scientific knowledge, chemical and mineralogical data became available along with petrographic data which led to classification based on rock genesis, mineral and chemical compositions, textural characteristics and mode of occurrence. The purpose of classification may vary as per the purpose viz. needs of research, comparison, teaching, field work and rock genesis. In the field, megascopic characters may prove useful while in the laboratory, chemical and mineral compositions would be better options for research. The latest classification of igneous rocks is based on three methods:

(1) Megascopic schemes: These are more useful in the field study and are based on the appearance of the rock in hand specimen or under the magnifying lens.

(2) Microscopic schemes: These are useful for identifying the individual minerals and the rock under the petrological microscope, where detailed information is required.

(3) Chemical schemes: These are mainly employed for research purposes, when in-depth information about the atomic and molecular behaviour of the minerals is essential.

For practical purposes, the minerals and the texture of the rock generally provide better information regarding the rock genesis as compared to the bulk chemical composition. The reason for this being that rocks like granite, rhyolite and obsidian all have the same chemical composition but they have a very different mineral assemblage and rock genesis. It is, therefore, difficult to classify them only on the basis of the megascopic, microscopic or chemical schemes. Granites are formed due to a slow rate of cooling at plutonic depth and under high pressure, while rhyolite and obsidian have crystallised at a rapid rate of cooling on the earth's surface at atmospheric pressure. Thus, igneous rocks show a great deal of chemical, mineralogical and textural variation which accounts for the general disagreement among the petrologists to classify them under one scheme. However, during the preliminary studies in igneous petrology, it would be useful to have a broad classification of igneous rocks based on the mineral composition, the depth of rock formation and the colour index of the rocks (Table 15.1).

15.2 CLASSIFICATION BASED ON MINERAL COMPOSITION

Quartz and the feldspar group of minerals are very useful in this regard. The presence or the absence of quartz in a rock decides the nomenclature of the rocks as acidic, intermediate and basic. Thus, acidic rocks contain abundant quartz (60% silica) and few Fe-Mg minerals (pyroxene and amphiboles). Basic rocks do not contain quartz, but dominantly contain pyroxenes, amphiboles and feldspars. The rock that possesses an intermediate amount of free silica and Fe-Mg minerals, as compared to the above two, is classified as an intermediate rock.

Feldspars have a widespread occurrence in the igneous rocks and are therefore, very useful in classifying them. For this purpose, feldspars are divided into two distinct groups, viz.

(1)   Alkali feldspars consisting of potassium and sodium feldspars such as orthoclase, microcline, sanidine (potassium feldspars) and albite (sodium feldspar).

(2)   Calc-alkali feldspars consisting of anorthite plagioclase which are essentially calcium feldspars.

15.3 CLASSIFICATION BASED ON DEPTH OF FORMATION

Igneous rocks which are formed at very great depths inside the earth are called plutonic rocks. They have a slow rate of cooling and are characterised by a coarse grained, equigranular texture. Granite is a typical plutonic rock which has an abundance of quartz and orthoclase. Hence, it is further classified as acidic and belonging to the alkali feldspar series, based on its mineral composition.

Rocks which solidify on the earth's surface are described as volcanic. They have erupted either from a volcano or as lavas through fissures. A very rapid rate of cooling of the lava makes these rocks fine grained or even glassy. They show characteristic features such as vesicles, amygdales and flow structure. Volcanic rocks show inequigranular textures under the microscope, e.g., basalt.

The hypabyssal rocks consolidate at depths which are intermediate between volcanic and plutonic types and they show characters which are intermediate between the plutonic and the volcanic rocks.

15.4 CLASSIFICATION BASED ON COLOUR INDEX

The colour of a rock depends on its mineral content. Minerals like quartz and alkali feldspars are colourless, white or pink, and they impart a light colour to the rock. The dark ferromagnesian minerals like augite, hornblende or olivine impart a dark colour to the rock.

Hence, the acidic rocks appear with a light shade (due to an abundance of quartz) and are called leucocratic. The basic rocks which appear dark (due to a significant presence of hornblende, olivine and augite) are called melanocratic. The intermediate rocks have a colour index somewhere in between the acidic and basic rocks and are called mesocratic.

15.5 DESCRIPTION OF IGNEOUS ROCKS

15.5.1 The Plutonic Rocks

The common examples of the plutonic rocks are granite, syenite, gabbro and dunite. These rocks show a holocrystalline and equigranular (granitic) texture.

Granite: Granites occur as huge plutonic bodies, commonly described as batholiths. Relatively smaller bodies are called as 'stocks' and 'bosses'. The common texture shown by granite is granitic or holocrystalline with an abundance of equidimensional grains of quartz, orthoclase or microcline with or without ferromagnesian minerals (mica, amphiboles or pyroxenes) and accessory minerals like zircon, rutile or apatite. The equigranular texture is formed by subhedral mineral grains, giving rise to a hypidiomorphic texture. Exceptionally, some granites may show porphyritic texture with phenocrysts of orthoclase or microcline embedded in an otherwise granitic texture. Megascopically or microscopically, the granites typically show equigranular texture. Based on the occurrence of certain minerals in the granite, it may be called as tourmaline, mica or hornblende granite.

Syenite: Syenite occurs as stocks, sills or dykes showing a holocrystalline, hypidiomorphic, granitic texture. Subhedral grains of quartz, orthoclase, microcline, augite or hornblende occur as coarse to medium sized grains. The amount of quartz, orthoclase, microcline or plagioclase is markedly less as compared to granite. Some low-silica syenites contain nepheline in place of albite. The mafic minerals may be biotite, augite or hornblende while the accessory minerals may be zircon, apatite and iron oxide.

Diorite: Diorite occurs as stocks, bosses or along the margin of granite batholiths. The rock is hypidiomorphic with coarse to medium sized grains showing a granitic texture. It has an abundance of plagioclase feldspars and little orthoclase or albite and quartz. The mafic mineral is mainly hornblende while biotite may or may not occur. The accessory minerals may be apatite, sphene, zircon and iron oxides.

Gabbro: The gabbroic intrusions vary in size from huge stocks and lopoliths to thick sills, dykes and plugs and are comparable in size to the granitic intrusions. The rock is holocrystalline coarse grained, hypidiomorphic, showing granitic texture. The dominant minerals are plagioclase, augite and olivine while hornblende and biotite are rare in gabbros.

Dunite: It is a mono-mineralic rock made up entirely of grains of olivine. As it is devoid of quartz or feldspars, it is classified as an ultrabasic rock. It is usually associated with gabbros and often with chromium and other ore deposits.

15.5.2 The Hypabyssal Rocks

These include pegmatite, dolerite and pitchstone. The pegmatites and dolerites are linear intrusions into the country rock at shallow depths. They are of variable extent and have a varied grain size. Pitchstone has a glassy texture.

Pegmatite: It is a very coarse grained intrusive rock which occurs in the form of dykes and veins. Quartz and orthoclase / microcline are the major minerals of the pegmatites which are also known for the occurrence of large grains of mica, beryl, tourmaline, topaz and fluorspar, besides important ores of tungsten, copper, tin, uranium and radium. The coarse grain size of the pegmatites is due to the end stage crystallisation of a granitic magma rich in volatiles.

Dolerite: This rock occurs as dark coloured intrusive dykes, veins and sills. The essential minerals are augite, hornblende, plagioclase and iron oxides. The chilled margins of dolerites may be glassy. The rock appears fine grained in megascopic specimens, but shows an inequigranular texture under the microscope.

Pitchstone: It is a dark, glassy rock with a dull pitch like appearance. It occurs as a part of a fast cooled lava flow. Under the microscope, it shows a holohyaline texture sometimes with traces of an embryonic growth of minerals.

15.5.3 The Volcanic Rocks

These include rhyolite, basalt and pumice.

Rhyolite: It occurs as lava and is characterised by flow structure both in hand specimens as well as under the microscope and is the volcanic equivalent of granite. Rhyolite shows a porphyritic texture in which phenocrysts of sanidine or orthoclase are embedded in a glassy or cryptocrystalline groundmass.

Basalt: The basalt rocks occur on the earth's surface as extensive lava flows of considerable dimension. They are the most abundant of all the volcanic rock types and contain augite, plagioclase and iron oxides as their major minerals. The rocks are fine grained showing a porphyritic texture and also vesicular, amygdaloidal and flow structures.

Pumice: This is such a light rock that it floats on water. It is the froth of the lava which has been explosively ejected from a volcano. The large number of air cavities in it renders it very porous and therefore, it is very light. The volume of its pore spaces may even exceed the glass in the rock. Commercially it is sold under the name of soap-stone.

The tabular classification of these rocks has been shown in the table 15.1.

Table 15.1: Tabular classification of igneous rocks

Type	Plutonic		Hypabyssal		Volcanic	
	Alkali	Calc-alk.	Alkali	Calc-alk	Alkali	Calc-alk
Acidic	Granite		Pitchstone Pegmatite		Obsidian Rhyolite Pumice	
Intermediate	Syenite	Diorite			Trachyte	Andesite
Basic		Gabbro		Dolerite		Basalt
Ultra-basic	Dunite				Limburgite	

POINTS TO REMEMBER

- The classification of igneous rocks is based on three methods chemical schemes, microscopic schemes, megascopic schemes.
- Megascopic schemes are useful in the field study and are based on the appearance of the rock in hand specimen or under the magnifying lens.
- Microscopic schemes are useful for identifying the individual minerals and the rock under the petrological microscope.
- Chemical schemes are employed for research purposes, when in-depth information about the atomic and molecular behaviour of the minerals is essential.
- The presence or the absence of quartz in a rock decides the nomenclature of the rocks as acidic, intermediate and basic.
- Acidic rocks contain abundant quartz (60% silica) and few Fe-Mg minerals (pyroxene and amphiboles).
- Basic rocks do not contain quartz, but dominantly contain pyroxenes, amphiboles and feldspars.

- The rock that possesses an intermediate amount of free silica and Fe-Mg minerals, as compared to the above two, is classified as intermediate rock.
- Igneous rocks which formed at very great depths inside the earth are called plutonic rocks.
- Rocks which solidify on the earth's surface are described as volcanic.
- The hypabyssal rocks consolidate at depths which are intermediate between volcanic and plutonic types and they show characters which are intermediate between the plutonic and the volcanic rocks.
- The Plutonic rocks show a holocrystalline and equigranular (granitic) texture.
- The common examples of the plutonic rocks are granite, syenite, gabbro and dunite.
- The acidic rocks appear with a light shade due to an abundance of quartz and are called leucocratic.
- The basic rocks appear dark due to a significant presence of hornblende, olivine and augite and are called melanocratic.
- The intermediate rocks have a colour index in between the acidic and basic rocks and are called mesocratic.

EXERCISE

1. Describe the classification of igneous rocks based on mineral composition.
2. Give the tabular classification of the igneous rocks and describe any two rocks in detail.
3. Write notes on:
   (a) Granite
   (b) Basalt
   (c) Dunite
   (d) Rhyolite
   (e) Pumice
   (f) Dolerite

Chapter **16**...

SEDIMENTARY PETROLOGY

Contents ...

16.1 INTRODUCTION

Secondary rocks are one of the three major groups of rocks that make up the earth's crust, the other two being the igneous and metamorphic rocks. Secondary rocks are derived from the breaking down and weathering of the pre-existing rocks on the earth's surface. These rocks cover about 66% of the earth's surface, including the continental masses and the ocean floor and their average thickness is about 2 kms for the continental crust and 1 km for the ocean floor. However, they occur as a thin veneer over the earth's solidified crust, accounting for only 5% of the volume of the crust.

Most secondary rocks are formed from the loose fragments which have been derived by the weathering of the pre-existing rocks.

In 1932, **H. A. Wadell** coined the term **sedimentology** for the scientific study of sediments. Geologists **James Hutton** and **Sir Charles Lyell** were pioneers in the study of sedimentology. They summarised that the sediments deposited in the geological past (millions of years ago) were formed by the geological processes, similar to those in operation today under the prevailing climatic conditions. This is an important assumption which is known as the **principle of uniformitarianism**. It states that 'the present is a key to the past'. Thus, the geological processes of nature have remained the same even though the energies of the natural agents like water and wind may have changed over the geological past. Therefore, by measuring the thickness of a sediment and assuming the rate of its deposition, it is possible to calculate the time required for its deposition.

16.2 SEDIMENT

Rocks which occur on the earth's surface are subjected to the vagaries of climate and to the action of the earth's atmosphere and the hydrosphere. This causes a breaking down of the rock by mechanical disintegration and chemical decomposition. This breaking down process of the rocks is called as weathering and it gives rise to the products of weathering i.e., the formation of loose rock material of varying sizes. These weathered debris are then transported by rivers, wind or glaciers as suspended particles or as material in solution which is later deposited in a basin. The consolidation of this loose material gives rise to the formation of sedimentary rocks on compaction and cementation. The word **sediment** implies that the solid material has settled down from a state of suspension either in air or water. However, in geological terms the word sediment also applies to the rock matter which has settled from suspension not only from water but also from glaciers, wind and chemical precipitates. The whole process takes place at surface conditions, i.e., at normal temperature and pressure.

Sedimentation:

In geology, the process of deposition of a solid material from a state of suspension or solution in a fluid (usually air or water) is termed as **sedimentation**. It also includes deposits from glacial ice and materials collected under the effect of gravity alone or accumulations of rock debris at the base of cliffs. Hence, the process of sedimentation includes all those surficial geological processes which bring about the formation of sedimentary rocks. It also comprises of studies regarding origin, transportation and deposition of the rock-forming materials, along with their diagenesis and lithification. The description, classification and interpretation of sedimentary rocks are also a part of this exercise.

It is with the help of studies related to processes of sedimentation and palaeontology that the stratigraphy of any region can be established. These studies have also proved useful in understanding the physico-chemical and bio-chemical behaviour of sedimentary materials during transport and post-depositional phases. In addition, the research in the environments of sedimentation has been instrumental in unfolding the mysteries of ancient sedimentary rocks and their geological environments.

16.3 SEDIMENTARY ENVIRONMENT

Definition : The sedimentary environment is a complex of physical, chemical and biological conditions under which a sediment accumulates. Sedimentary environments are either of erosional or depositional type and at the same time, they may be formed during the geological past or as present-day accumulation.

Types of sedimentary environments:

Sedimentation at any place is controlled by the varied relationship between the geological processes which operate on land or sea; and due to geological agents like water, ice, wind, waves or currents. It is the variation in these processes that the beach sediments differ from that of the flood-plain deposits or sand dunes.

In the present text, only the depositional environments of the present day have been discussed. The present-day sedimentary environments are studied under 2 major heads, namely (1) the physical, chemical and biological processes that cause sedimentation and (2) the kinds of sedimentary deposits.

Due to the complexity of its formation, the sedimentary environment may be classified using different criteria or bases, depending upon the importance given to certain sedimentological features. For example, the physico-chemical classification is based on certain environmental factors or elements. At the same time, the nature of depositional media (i.e., air, water, ice or wind), or water depth alone may be used to classify the marine environment.

Twenhofel in 1950, has classified the sedimentary environments into three major types namely, the continental, marine and transitional as shown in Table 16.1.

Table 16.1 : Classification of sedimentary environments
(after Twenhofel, 1950)

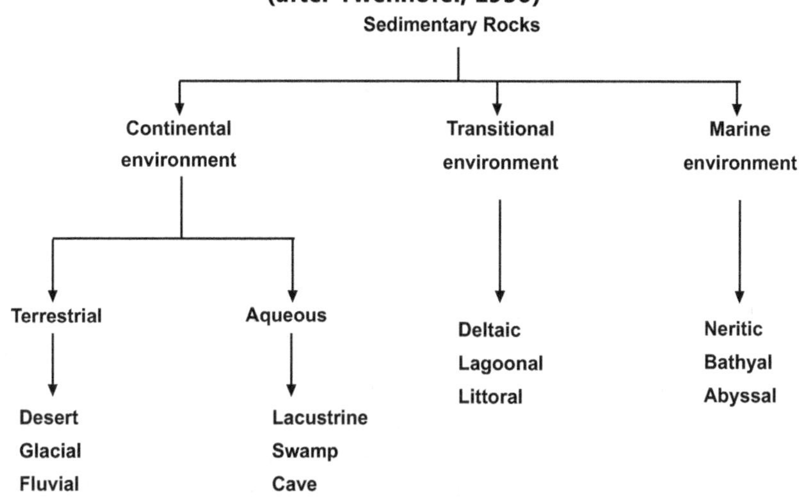

1. Continental or non-marine environment:

These are sedimentary environments where the surface of deposition normally lies above the sea level (with exception of Death Valley). These are further sub-classified as either **terrestrial** or **aqueous**. In terrestrial environment, the deposits are laid down either by wind or glaciers. The desertic environment allows formation of sand dunes. In glacial environment, the sediments are carried, in suspension by glacial ice and deposited in suitable places forming moraines and varves. On the other hand, the aqueous (or fluvial) environment includes those deposits formed by sediments transported in suspension or solution by running water and deposited in suitable basins, e.g., lakes, river banks and flood plains. The saline deposits such as Great Salt Lake are also included as aqueous deposits. A special category of alluvial environment has been included for the deposits formed in river channels and associated flood plains. Lacustrine sediments accumulate when glacial lakes are formed and allow sedimentation. Swamps are developed when sea water accumulates along the shore lines and forms a depositionary site. The percolating groundwater may also deposit some of its dissolved constituents, forming the cave deposits.

2. Transitional environment:

The depositional features which are intermediate between continental and marine types are classified as transitional environments. The delta is a sedimentary deposit fed by a stream, and distributed by waves and currents of lake and sea. A lagoonal deposit is a shallow water sediment formed along the barrier beaches, spits or bars along the near-shore zones of the sea.

Littoral environment extends from the region of high tide to that of low tidal zones of the sea (e.g., beach deposits).

3. Marine environment:

The marine depositionary basins are dominated by the formation of mainly limestones but different facies of shallow and deep marine sediments have also been identified. Various types of chemical and organic limestones are characteristic of marine deposition.

The neritic zone of the marine environment extends between low tide level to a depth of 600 feet. The bathyal environment, however, lies between 600 to 6000 feet. While, the abyssal zone comprises of deepest portion of the sea beyond 6000 feet. A special category of hadal environment has been assigned for the deposits formed at depths greater than 21000 feet.

16.4 FORMATION OF SEDIMENTARY ROCKS

The main source for the sediments is the weathered material formed on the surface. It is the process of weathering, erosion and denudation that these loosened debris are readily made available for the transport. It is important to differentiate between the processes of weathering, erosion and denudation, as these terms are repeatedly used to describe the geological work of the natural agencies on the earth's surface. **Weathering** involves the physical and chemical destruction of rocks *in situ*, i.e., at the place of its origin and its weathering products accumulate on the rock itself.

Erosion is the 'wearing away' of the rocks and the land surface. The weathered rocks are easily etched or dislodged by gravity, running water and wind. Thus, the process of the breaking down of the rocks and the transport of the resulting product from the place of its formation is called as **erosion**.

Due to the erosion, the underlying rocks get exposed and are subjected to the process of weathering. Thus, the process of removal of the weathered rock and exposure of the fresh country rock to further erosion is termed as **denudation**.

In nature, these three processes form a cycle one after the other. Thus, weathering leads to erosion, erosion further gives rise to denudation and at the end of the denudational cycle, the next cycle of weathering commences again, and the processes go on.

The rocks of the earth's surface are prone to the vagaries of climate and also subject to the action of natural agencies like running water, underground water, blowing wind and movement of glaciers, all of which are continuous and ceaseless operations. These agencies bring about a continuous wear and tear of the rock, causing them to break down by mechanical disintegration and chemical decomposition. As stated earlier, during physical weathering, the rock disintegrates by the action of water, wind, ice and a wide variation in diurnal temperature; while in the case of chemical weathering, the minerals in the rocks are partially dissolved or altered by the water, leaving behind an insoluble residue. The disintegrated rock consists of broken fragments of varying sizes and a chemically decomposed, fine grained insoluble residue of different compositions. The weathered products covering the rock are called as the **regolith**. It may remain at the place of its formation indefinitely or may, over a period of time, be transported to its ultimate destination, the sea.

Hence, the material may not always be transported and could remain in place as a weathered rock, such as a laterite or as a loose mantle on the rocks, such as soil. These weathered rocks or soils which have suffered little or no transport, are described as residual deposits.

The source material for the formation of secondary rocks is thus derived from pre-existing igneous, metamorphic or even secondary rocks, which have been subjected to weathering at the earth's surface. Thus, it is clear that the loose material (weathering product) may either remain in place (in-situ) and be termed as a residual deposit, or it may be transported into a basin.

The loose material is transported either as particles in suspension or as material in solution. It gets deposited in hollows or basins, on the land or in the seas and oceans. The suspended material gets deposited wherever the velocity of the transporting media is reduced or checked or wherever there is a change in the physical conditions. The material in solution, precipitates from the solution due to a change in the physical or chemical nature of the media or indirectly by the activities of animals and plants.

16.5 TRANSPORTATION

The fragments and insoluble matter produced by weathering, may accumulate in situ or it may get transported along with the soluble matter, to be deposited elsewhere. Transport takes place mainly by moving waters, such as rivers, sea waves, ocean currents and by wind and glaciers. Rivers and wind are by far the most efficient means of transport for the weathered products. Rivers carry the material in suspension or in solution while the larger fragments are rolled along its bed. The material in solution moves either as ground water or as surface run off, until it is discharged into the streams and rivers and carried to the sea. Thus, the grain size of the material in suspension ranges from the coarsest to the finest and along with the material in solution, it is transported with several possible halts along the way, to be deposited finally into the sea. Transport by sea waves and currents is most effective along the coasts. Wind is perhaps the most potent transporting agent, as it carries fine deposits over very large distances. The best example is the loess of China, a deposit which has been blown there by winds all the way from the Asian deserts. Sand dunes are other examples of the transporting power of wind. Glaciers carry loose fragments of rock, of all possible sizes, on its surface or within its frozen body or even dragged below it. Its transporting capacity is revealed in the great thickness and the large aerial extent of glacial deposits like boulder-clay and moraines, deposited in India during the Pleistocene glaciation.

The effect of transport on the rock fragments carried in suspension, is to break them into smaller fragments and particles, with a rounding of their angular edges. Longer the distance of transport, greater is the rounding of the grains. Amongst the transporting agent, wind is the most efficient in rounding the transported grains and produces near spherical grains. Rivers, sea and ocean waves are also quite effective agents for rounding of grains. Glacial transport, however, due to its solid state and low mobility, does not bring about much rounding of the grains. Transporting agents also cause the sorting of the grains. Sorting involves segregation of the grains as per their size and density. This again depends upon the mobility of the transporting agent; thus greater the mobility, greater is the sorting. It also involves the separation of coarse, medium and fine particles into definite zones along a beach, a river bed or in a sand dune. Wind is again the most efficient agent of sorting while glaciers are the least efficient.

16.6 DEPOSITION

The final destination of all the material transported by water, wind or ice, is the sea. On the way, the material may be deposited in basins on the land, where they may accumulate over geological time and give rise to thick, fresh water deposits, known as the continental deposits. Most of the material, however, is transported to the sea where it is deposited and gives rise to the marine deposits.

The actual deposition may be effected either by mechanical or chemical means, depending upon whether the transported material is being carried in suspension or in solution. Suspended material gets deposited whenever the transporting agent (water, wind and glacier) gets overloaded, or its velocity is checked or if there is a chemical or physical change. At the mouth of large rivers, the velocity of the river water is checked as it meets the sea and this results in the deposition of the suspended matter in the shape of a delta. Deposition also takes place by a process of flocculation, as the saline sea water gets mixed with the deposit carrying fresh water of the rivers.

The soluble matter in the water gets deposited on the land or in the sea by physico-chemical processes like precipitation or evaporation or indirectly by the activities of organisms. Chemical reactions may take place between the water containing dissolved matter when it meets water containing different substances, bringing about a precipitation of the dissolved matter. Deposition can take place due to the evaporation of water from a solution which has become oversaturated. A large amount of matter in solution reaches the sea, where its accumulation over a long geological time, has contributed to the present salinity of the sea and the oceans.

Some living organisms extract salts of calcium and magnesium from the matter in solution, in order to build their shells and skeletons. When these organisms die, their hard parts accumulate to form extensive deposits. The activity of bacteria is effective in the deposition of bog-iron ore in swamps and lakes, while algae are responsible for the formation of some calcareous deposits.

16.7 COMPACTION, CEMENTATION AND LITHIFICATION

In the final phases of formation of sedimentary rocks, the deposited sediments get compacted and cemented during the process called lithification. During this process, the soft sediments become rigid rock (*lithi* means rock in Greek). It begins when the sediment is laid down and is gradually buried. It then gets compressed under the burden of overlying new sediment cover. A recently deposited sediment, therefore, is usually a loose material, with open spaces or pores filled with air or water. The process of lithification acts to reduce that pore space and replace it with solid mineral material. The main processes involved in lithification are compaction and cementation. Compaction involves squeezing of the sediments into a smaller volume by packing them into more close spaces, by removing water from the pore space (desiccation) or by pressure solution at the points where sediment grains contact each other.

Simultaneously, cementation involves filling up of the pore spaces with solid minerals (usually calcite or quartz) that are deposited from solution or that enable existing sediment grains to grow into the pores. The pore spaces may or may not be eliminated during lithification. Hence, at the end of the compaction and cementation, a sediment becomes rigid and transforms into a solid rock. This is then termed as a sedimentary rock.

16.8 CLASSIFICATION AND DESCRIPTION OF SEDIMENTARY ROCKS

By considering the manner in which the products of weathering are distributed transported and deposited in the course of geological time and their final consolidation into hard rocks, a simple classification for secondary rocks, based upon the products of weathering, is given in Table 16.2.

Table 16.2 : Classification of secondary rocks products of weathering

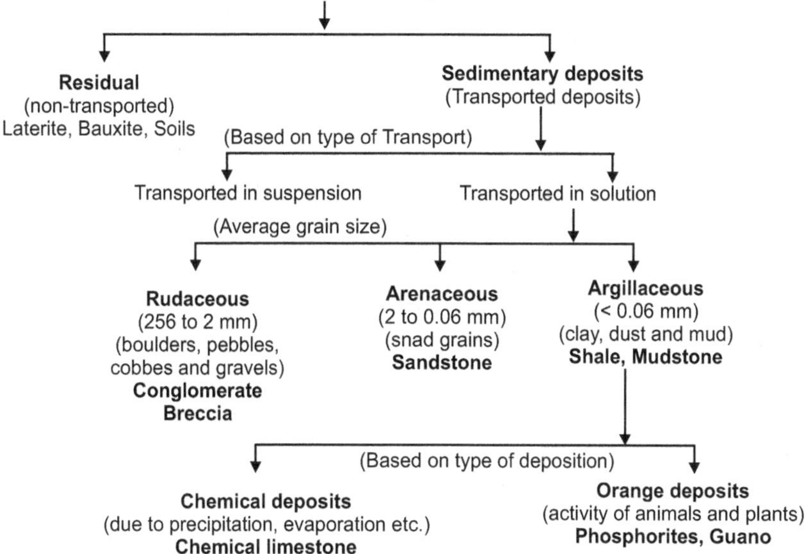

As explained earlier, the loose rock material which goes to form secondary rocks, is derived by the weathering of the pre-existing rocks on the earth's surface. The phenomena of weathering of rocks is brought about by two complementary processes viz.

i) Mechanical or physical disintegration and

ii) Chemical decomposition of the rocks.

The intensity of these two processes is a function of the climate, the temperature variation and the geological agent or agents in operation, on the rocks, in a certain area. Weathering, in effect, means the breaking down of the rock into smaller fragments and also of matter, which goes into solution.

In mechanical or physical disintegration, the rock is broken up without any chemical change. The breaking down is effected by the action of (a) a large variation in day and night temperatures, (b) frost action and (c) abrasion by water, ice or sand laden-winds, producing rock fragments of variable sizes.

In chemical decomposition, the minerals of rock are attacked by water and natural chemicals bringing about chemical changes. Some part of the mineral may go into solution and gets transported, while the insoluble part of the mineral remains in place as a residue. Usually the two processes of weathering act together, though one of them may be predominant. The combined effect of weathering is to break down the rocks into finer and more durable material, part of which goes into solution. This weathered material, which consists of broken fragments and a fine grained insoluble residue, of varying thickness and composition, covers the weathered rock as a mantle called as the regolith. The regolith may remain in place, or may over a period of time, be transported to its ultimate resting place in the sea.

16.8.1 Residual Rocks

Laterite: Laterite is a reddish porous, concretionary material, which covers vast areas in tropical and subtropical regions, forming a hard crust on the base rock. It is formed by the intensive weathering under tropical climatic conditions where wet and dry seasons alternate. Laterite varies greatly in composition, but in general, it consists of a mixture of hydrated iron oxide (ferric oxide) with hydroxides of alumina in varying proportions. The name is derived from *Later* (Latin for a brick). It is very soft when fresh but on exposure to the atmosphere; it becomes hard which makes it very useful in the manufacture of bricks. The colour of laterite also varies with its iron content. Thus, iron-rich laterite is reddish in colour but assumes yellow shade with increase in alumina content. It may show oolitic or pisolitic structure.

Bauxite : Bauxite is also a product of intensive weathering under alternate wet and dry conditions typical of tropical climates. It consists of the hydroxides of aluminium with iron oxides in lesser proportion. It usually has a buff colour but reddish tinge may be seen. Generally the texture of the rock is concretionary but at times, it may develop oolitic texture.

16.8.2 Rudaceous Rocks

Breccia : These rocks are made up of very coarse fragments, such as pebbles and cobbles (particle size between 10-200 mm). They are composed of hard, tough and durable material, which is resistant to weathering (e.g., minerals like quartz, jasper and chert). These fragments are cemented in either siliceous, ferruginous or calcareous cement. As the fragments have not undergone much transport before deposition, they remain angular. They are either formed from rocks eroded and rolled down from the mountain slopes or due to mechanical disintegration. However, the term 'volcanic breccia' is also used for the accumulation of angular fragments of volcanic origin.

Conglomerate: These are also large sized fragments like the breccias with the difference that they are rounded. They are formed due to the torrential action of a youthful river, where due to the friction along the river bed and prolonged transport, the angularity of the fragments becomes rounded or spherical in nature. As less resistant material gets disintegrated during long transport, only durable material survives in the form of detrital grains of conglomerate. These fragments are cemented in either siliceous, ferruginous or calcareous cement.

16.8.3 Arenaceous Rocks

Sandstones: Arenaceous rocks include the rocks which are formed by the consolidation of sand size particles (size between 0.1 to 2.0 mm, hence, called as 'sands'). These detrital fragments may comprise of quartz, chert, agate and varieties of feldspars. Some heavy minerals like garnet, zircon and tourmaline may also occur. Sands are classified as marine (sea), estuarine (creeks), lacustrine (lakes), fluviatile (river), desert, fluvio-glacial or volcanic according to their place of deposition. Those sediments which have been deposited under the influence of water are classified as 'aqueous sands'. The other types of sand are 'eoline' (wind blown) sands, which include desert and coastal dune sands.

When the loose sands are cemented together they give rise to sandstone, which is a hard and consolidated rock. As sandstones are formed by cementation of any kind of sands, they may also be classified as marine, fluviatile or desert sandstones. During consolidation, the particles get bound together by the deposition of cementing substance, including the detrital grains. The cement may be siliceous, ferruginous or calcareous.

Types of sandstones:

A coarse sandstone with angular grains is known as 'grit' while a sandstone with an abundance of feldspars, usually derived from the disintegration of granite, is termed as an arkose. Some sandstone, with considerable calcareous cement, is known as flagstones. Freestone is uniform, thick-bedded sandstone with few divisional planes, which enables it to be cut easily in any direction. It, therefore, forms a good building stone. Depending on the environment of transportation and deposition, sandstones may show different structures like ripple marks, graded bedding and current bedding.

16.8.4 Argillaceous Rocks

Mudstone: These are very fine grained rocks (average grain size is less than 0.01 mm) which form an important group of sedimentary rocks. They often exhibit perfect lamination and are earthy in texture. Many lake deposits and river alluvium contain large quantities of mudstones. After compaction and cementation, this fine grained material forms a hard compact rock. These rocks do not have any laminations and hence, are conveniently used for clay modelling. Due to the fine grained nature of these rocks, it is difficult to identify their component minerals with the naked eye. However, microscopic examination suggests that they are of micaceous and clayey composition.

Shale: It is an argillaceous rock, which is well bedded and which splits easily along its bedding planes. This property results from the presence of fine micaceous material in it. When the fine grained material is rich in carbonaceous material, the rock appears blackish and is then called as carbonaceous shale. On the other hand, when ferruginous or siliceous material dominates, the rock is called as ferruginous or siliceous shale, respectively.

Argillaceous rocks are also classified on the basis of their depositional origin as marine, aqueous, glacial or volcanic. Marl is a clay rock which contains an appreciable proportion of carbonates of lime and magnesium.

16.8.5 Chemical Deposits

Precipitation is a process which involves separation of the solute from the solution. The water contains ions of the dissolved substances as a result of rock weathering and these ions are carried away in solution and later on deposited directly by physico-chemical processes, such as evaporation, cooling of the solution, subsequent saturation of substances and precipitation. These rocks are fine grained, crystalline, although some evaporites may form large crystals. Based on their chemical compositions the deposits are classified as siliceous, carbonates, ferruginous deposits or salts.

16.8.6 Organic Deposits

These are sedimentary deposits, which are formed by the direct or indirect action of animals and plants. They are mostly fine grained rocks. The materials mainly accumulate on the sea floor, but some fresh water and terrestrial examples are also known. The deposition may build up directly through organisms such as a coral rock.

Biochemical deposits are favoured by the activities of an organism which promotes precipitation (e.g., algal limestone).

Biomechanical deposits are those which are formed by the accumulation and consolidation of organic material (e.g., gastropod limestone). The most common precipitate is that of calcium carbonate of the shells, but silica also gets precipitated by the activity of radiolarians, diatoms and other microscopic organisms and plants. Faecal pellets are the excreta of the marine invertebrates.

The organic deposits are also classified on the basis of their chemical composition and some of the important organic deposits are listed below:

1. Calcareous   Limestone
2. Phosphatic   Phosphorites, Guano (accumulation of birds' excreta)
3. Ferruginous   Bacterial iron ore
4. Siliceous   Radiolarian and diatom oozes
5. Carbonaceous   Coal and its varieties

POINTS TO REMEMBER

- The principle of uniformitarianism states that 'the present is a key to the past'.
- The process of deposition of a solid material from a state of suspension or solution in a fluid (usually air or water) is termed as sedimentation.
- The sedimentary environment is a complex of physical, chemical and biological conditions under which a sediment accumulates.
- Continental or non-marine environment are sedimentary environments where the surface of deposition normally lies above the sea level.
- Transitional environment comprises of depositional features which are intermediate between continental and marine types
- The process of the breaking down of the rocks and the transport of the resulting product from the place of its formation is called as erosion.
- The process of removal of the weathered rock and exposure of the fresh country rock to further erosion is termed as denudation.
- The weathered products covering the rock are called as the regolith.
- The final destination of all the material transported by water, wind or ice, is the sea.
- In the final phases of formation of sedimentary rocks, the deposited sediments get compacted and cemented during the process called lithification.
- Weathering involves the physical and chemical destruction of rocks *in situ*, and its weathering products accumulate on the rock itself.
- Erosion is the 'wearing away' of the rocks and the land surface.
- the process of removal of the weathered rock and exposure of the fresh country rock to further erosion is termed as denudation.
- Examples of residual rocks are laterite and bauxite.
- Examples of rudaceous rocks are breccia and conglomerate.
- Examples of arenaceous rocks are various types of sandstones.
- Examples of argillaceous rocks are mudstone and shale.
- Examples of chemical deposits are evaporites and salts.
- Examples of organic deposits are limestone, phosphorites, guano, coal, etc.

EXERCISE

1. What is meant by the term 'sedimentary environment'? How are these classified? Explain one of them in detail.

2. Write notes on

   (a) Continental environments

   (b) Marine environments

   (c) Transitional environments

   (d) Weathering, erosion and denudation

   (e) Transportation and deposition of sediments

   (f) Derivation of secondary rocks

   (g) Lithification process

   (h) Residual rocks

   (I) Arenaceous rocks

   (j) Argillaceous rocks

   (k) Chemical deposits

Chapter 17

TEXTURES AND PRIMARY STRUCTURES OF SEDIMENTARY ROCKS

Contents ...

17.1 INTRODUCTION

Two kinds of textures are recognised in secondary rocks, viz.

(i) clastic and (ii) non-clastic

17.1.1 Clastic Texture

Secondary rocks formed by the accumulation of rock fragments or minerals show clastic texture. The loosely or lightly packed fragments or particles may be of any size, form or even composition. Detrital or clastic aggregates such as gravel or sand have some intergranular porosity. Finer particles occur between the larger particles and serve as a bond or matrix. The matrix is more appropriately called cement, when it results from the crystallisation of minerals between the grains. All the detrital rocks show clastic texture. Non-detrital sediments occurring on the sea floor, such as shell fragments of animals also show clastic texture.

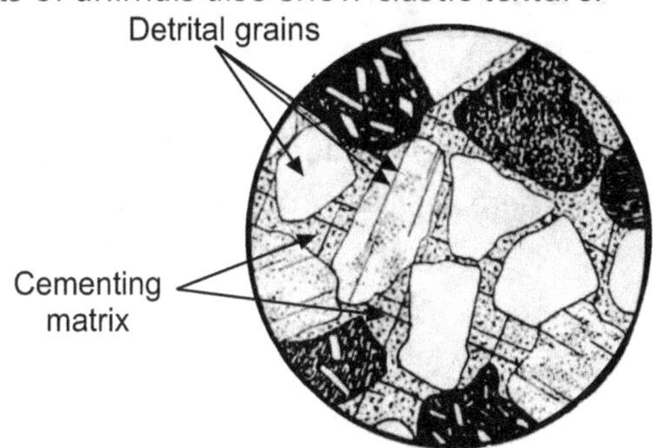

Fig. 17.1: Clastic texture

Examples: Sandstone, mudstone and shell limestone.

17.1.2 Non-clastic Texture

This texture is described by the interlocking or mosaic pattern shown by the crystals in the rock. The texture is often referred to as crystalline. The individual crystals or grains may show a variety of shapes and sizes and the mineral boundaries may be irregular or smooth. Such a texture, where the mineral grains are identifiable only under the microscope, is described as micro-crystalline. This texture results from the deposition and growth of crystals within an aggregate by precipitation from solutions. These crystals may recrystallise during diagenesis.

Fig. 17.2: Non-clastic texture

Examples: Coralline limestone, stalactites, stalagmites and calctufa.

Grain size classification of the detrital sediments

Detrital sediments are classified according to their grain size into rudaceous, arenaceous and argillaceous sediments.

Rudaceous sediments have a grain size between 200 mm to 10 mm and consist of boulders, cobbles and pebbles, which may be angular or rounded in shape. Angular grains in a rudaceous rock give rise to breccias, whereas the rounded grains give rise to conglomerates.

Arenaceous sediments contain coarse, medium to fine sand grains having a size between 2 mm to 0.1 mm giving rise to sandstones.

Argillaceous sediments are very fine grained and consist of mud, silt, clay and dust, having a grain size less than 0.1 mm. They form the shales, mudstones, siltstones and clays.

17.2 STRUCTURES OF SEDIMENTARY ROCKS

The peculiar mode of deposition of detrital sediments gives rise to various structures, which are characteristic of sedimentary rocks. Most typical is the stratification of the rocks, the graded and current bedding, ripple marks, mud cracks; rain prints and animal tracks and trails.

17.2.1 Lamination and Stratification

Lamination : Sedimentary rocks get deposited in layers, beds or strata. Lamination is a small-scale sequence of fine layers that occurs in **sedimentary rocks**. Laminae are normally smaller and less pronounced than bedding. A single **sedimentary rock** can have both laminae and beds.

The stratification is recognised by the difference in composition, texture and colour of the strata which are arranged in roughly parallel bands (Fig. 17.3).

These bands, which also vary in hardness and compaction, are brought out more prominently when the rock undergoes weathering. The plane of junction between beds is called the **bedding plane**. A single strata bounded on both sides by bedding planes is called a **stratum** or **bed**. The thickness of a bed may vary from a few centimetres to several metres. Clay or mud sized particles give rise to very thin beds known as **laminae**, which split easily, due to the presence of micas.

Fig. 17.3: Stratification in sedimentary rocks

17.2.2 Bedding

The sedimentary rocks are usually laid down in continuous and parallel horizontal layers. But, when these layers are deposited along hill slopes, then the debris are deposited making an inclination with the surface of deposition. So a sequence of sedimentary rocks deposited as scree contains layering that is inclined to the surface and is said to have a depositional dip.

Therefore, when depositional layers are parallel to one another, then it is termed as concordant bedding. It indicates stability conditions during deposition. On the other hand, when bedding planes are not parallel to one another, then it is termed as discordant bedding. It is an indication of rapid changes in the direction and strength of the sediment carrying stream or wind.

17.2.3 Graded Bedding

When loose detrital matter is deposited in a still water basin, the larger fragments tend to settle down at the bottom of the basin. These are followed upwards by the smaller fragments and finally by the finest particles right to the top (Fig. 17.4).

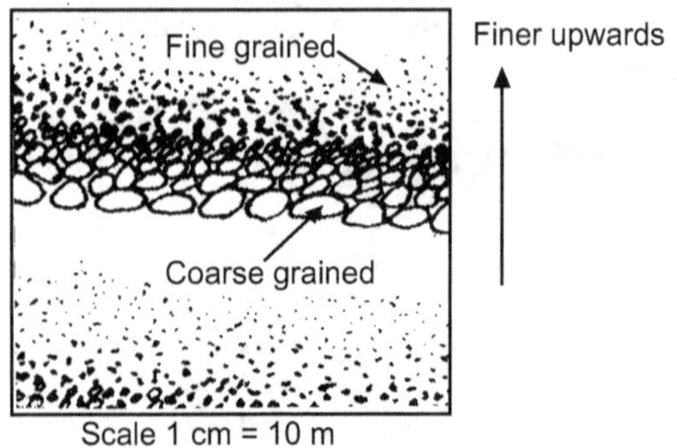

Fig. 17.4: Graded bedding in sedimentary rocks

This gives rise to a regular size wise stratification of the detrital matter, which is called graded bedding.

17.2.4 Current Bedding or Cross Bedding

Bedding planes are often approximately parallel to each other and the bedding is said to be concordant. However, sometimes within the beds, a subsidiary stratification is seen, which is inclined to the main bedding planes. This stratification is, therefore, discordant to the bedding planes and is variously, called as cross, current or false-bedding. Current or cross bedding is an indication of rapid changes in the direction and strength of the sediment carrying stream or wind. This structure is especially well seen in the deltas of large rivers. A more pronounced cross bedding called as torrential bedding is seen when coarse deposits are laid down alternately with fine horizontal laminae, during alternate floods and quiet periods. It is also seen in alluvial fans in semi-arid regions. Aeolian current bedding, brought about by wind, shows curved cross lamination on a larger scale. This structure is commonly found in sand and silt deposits.

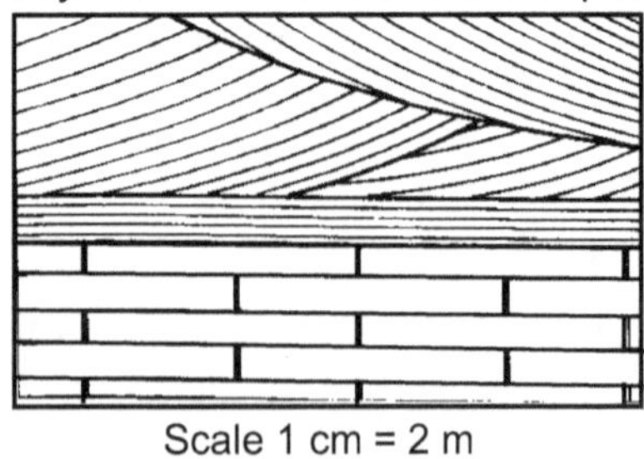

Fig. 17.5: Current or Cross bedding

17.2.5 Ripple Marks

It is an easily recognisable wavy pattern observed on the surface of a sedimentary rock.

It is similar to the wavy pattern seen in beach sands and which has been preserved by the sediments, under special conditions.

On beaches the ripple marks are produced by wave actions and the cross section of the ripple marks are symmetrical. Asymmetrical cross-sections of ripple marks indicate current action either in water or in the air.

Fig. 17.6: Ripple marks

Ripple marks formed by water and those formed by wind can be easily distinguished. In aeolian ripple marks, the coarser grains are found on the crest and the finer grains in the troughs of the waves. In the case of water formed ripple-marks, however, the finer grains are on the crests and the coarser ones are in the troughs. Ripple marks are commonly seen in sandstones.

17.2.6 Mud Cracks

Mud cracks are seen on the floor of any dried up pool of water or on the dry river bed (Fig. 17.7).

Fig. 17.7: Mud cracks or sun cracks

These are preserved in fine grained sedimentary rocks like mudstones and clays. They form a network of polygonal fissures which have been filled in with sand or mud and preserved on the surface of the rock. Mud cracks indicate conditions of dehydration and drying of clayey sediments on their prolonged exposure to the atmosphere.

17.2.7 Rain Prints

These are the shallow depressions preserved on the surface of the sedimentary rock, caused by the impact of rain drops on the rock (Fig. 17.8).

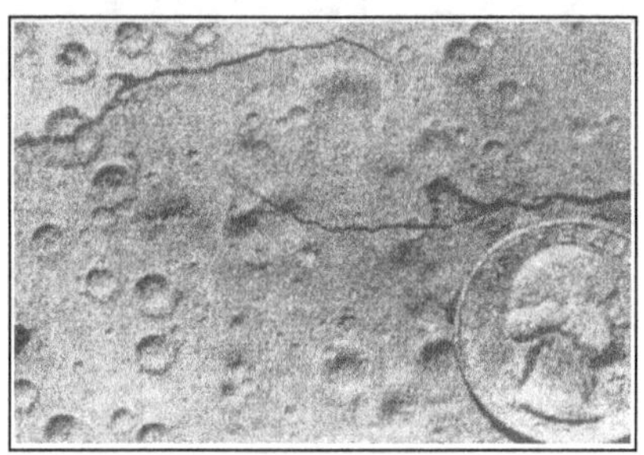

Fig. 17.8: Rain prints

The shallow depression is encircled by a low ridge. For obliquely falling rain drops, the ridge appears higher on the lee side as compared to that on the windward side.

POINTS TO REMEMBER

- There are two textures in secondary rocks- clastic and non-clastic.
- Secondary rocks formed by the accumulation of rock fragments or minerals show clastic texture.
- Non-clastic texture is often referred to as crystalline.
- It is described by the interlocking or mosaic pattern shown by the crystals in the rock.
- Stratification is the deposition of sedimentary rocks in layers, beds or strata.
- When loose detrital matter is deposited in a still water basin, the larger fragments tend to settle down at the bottom of the basin which gives rise to a regular size wise stratification of the detrital matter, known as the graded bedding.

- Ripple marks are an easily recognisable wavy pattern observed on the surface of a sedimentary rock.
- Mud cracks are seen on the floor of any dried up pool of water or on the dry river bed.
- These are the shallow depressions preserved on the surface of the sedimentary rock, caused by the impact of rain drops on the rock
- When animals move on soft sediments, they leave behind markings according to their movements called tracks and trails.

EXERCISE

1. Define the term 'texture' of a sedimentary rock. Describe any two textures shown by a sedimentary rock.
2. Define and describe the following with the help of neat diagrams.

    (a) Stratification　　　　　(b) Graded bedding

    (c) Current bedding　　　　(d) Ripple marks

    (e) Mud cracks.

Chapter 18

METAMORPHIC PETROLOGY

Contents ...

18.1 INTRODUCTION

The survey of the folded mountains like Himalayas reveals that the pre-existing rocks have been converted by the pronounced temperature and pressure changes into a totally new rock variety, viz the metamorphic rocks. Metamorphic rocks are particularly found in the ancient crustal masses of the earth, as well as, in the mountain chains of the earth such as the Alpine-Himalayan, Rockies, Andes and the Appalachians. **Sir Charles Lyell** (1833), who studied the structure of the Alps noticed that some of the fossiliferous beds were converted into crystalline rocks with a peculiar alignment of the newly developed minerals and therefore, coined the term metamorphism for these rocks (*meta*-change, *morph*-form) which literally means, 'change of form'. These changes may occur either at the near-surface conditions or at great depths beneath the earth's crust. Any pre-existing rock affected by marked changes in heat, pressure and chemical environment will finally give rise to a metamorphic rock.

Definition:

Turner and **Verhoogen** (1960) defined metamorphism as the 'mineralogical and structural adjustment of solid rocks, to physical and chemical conditions, which have been imposed at depths below the surface zones of weathering and cementation and which differ from the conditions under which the rocks in question originated'. In general, metamorphism means a partial or complete recrystallisation of the pre-existing rocks and the production of new structures. Therefore, metamorphic rocks result when the pre-existing rocks lose their stability due to the pronounced changes of temperature, pressure and chemical environments below the shell of weathering and have re-established stability by adjusting to the new environment. Changes due to the pulverisation and crushing of the rocks with or without the recrystallisation of new minerals are also termed as metamorphic effects. However, the chemical and physical changes that occur during the process of weathering, leaching and cementation by meteoric waters at near-surface conditions at normal temperature and pressures, are not considered as metamorphic changes.

Crystallisation of minerals takes place both in the igneous and the metamorphic rocks but there is a basic difference in the mode of crystallisation. The metamorphic reactions are initiated with increasing temperatures and take place only at the contacts of the mineral grains where the liquids are present, while the rest of the rock remains essentially in a solid state. Crystallisation in igneous rocks starts in a liquid phase and continues with a fall in temperature of the magma.

The metamorphic phenomenon takes place because of the progressive changes in the major factors of temperature, pressure and chemically active fluids.

18.2 AGENTS OF METAMORPHISM

Metamorphism results from the changes in the physical and chemical environment of the original rock. The main agents responsible for metamorphic changes are

(a)　Temperature

(b)　Pressure and

(c)　Presence of chemically active fluids.

In addition, the factors such as the amount of water present, the mineralogical composition of the original rock, the depth at which metamorphism occurs or the time span over which metamorphism takes place, are also important.

(a) Temperature plays a major role in the mechanical and the chemical processes of metamorphism. The rise in temperature usually takes place due to the proximity of magma or emplaced igneous rocks and also due to the ascension of gases and hot solutions from the earth's interior. The rise in temperature brings about recrystallisation of the minerals in the rock and accelerates the rate of chemical reactions such as dehydration and endothermic reactions within the rock. The increase of temperature renders most of the rocks ductile thereby increasing the speed of the geochemical reactions.

(2) During metamorphism, the pressure may act in one of the two ways, viz. i) as a **confining** or **lithostatic** pressure and ii) as a **directed** pressure (**stress**). The confining pressure is a non-directional pressure brought about by the weight of the overlying column of rocks above the place of metamorphism. With an increase in depth, the confining pressure increases correspondingly to an order of 250-300 bars per kilometre. Metamorphic changes usually occur at pressures upto ten thousand bars. The greatest effects of confining pressures on rocks are shown by those rocks which were buried during the great mountain building activity of the earth. Directed pressure on the other hand, is the most effective in the tectonically active zones of the lithosphere. It is also called as 'stress' within the rocks and is a non-uniform pressure. The effect of the directed pressure is to reduce the melting point of the minerals and increase their solubility. Thus, when the rocks are under stress, the recrystallisation starts at the points where the mineral contacts contain pore liquids which occupy the space between the minerals. The solid minerals in contact with the liquids then recrystallise perpendicular to the maximum stress direction (Fig.18.1). Hence, the directed pressures tend to change the shape of the original minerals and all this happens essentially in the solid state.

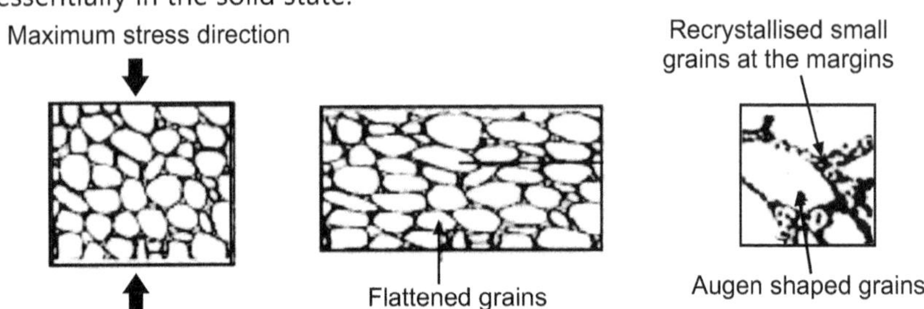

Fig. 18.1: Recrystallisation during metamorphic reactions

(3) The high temperature chemically active fluids, rising from the magma chamber, accelerate the metamorphic reactions by heating the rocks which come in contact with them. These fluids can be categorised as

    (1)   fluids originally present within the rocks,

    (2)   those formed during the chemical reactions in the rock and

    (3)   hot and chemically active fluids from the earth's interior.

During metamorphism, the chemical reactions attempt to restore the chemical equilibrium which had been disturbed under the new physico-chemical environment. The chemically active fluids not only initiate the chemical reactions but also help the new substances produced during the chemical reactions to migrate. They are thus responsible for the recrystallisation of minerals and the re-combination of the chemical constituents of the rock.

Water is the most potent fluid for effecting the migration of the products of the chemical reactions and it also serves as a catalyst for the various reactions. Moreover, the water molecules show no difference between their gaseous and liquid phases at higher temperatures. The fluid so formed is in a supercritical state and is able to penetrate the solid rocks through their minute pores and permeable zones. Laboratory experiments on rocks and their reactions reveal that at constant temperature, different mineral combinations can be produced by altering the partial pressure exerted by the water. Water, therefore, helps in the efficient transfer of chemicals and also in the crystallisation of metamorphic minerals.

The **composition** and **texture** of the original rock is also vital in working out the rate of the chemical reactions. Chemically susceptible rocks like limestone ($CaCO_3$) are more sensitive to metamorphism than rocks like gabbro and basalt (which contain iron and magnesium minerals). The fine-grained rocks are more prone to go into solution than the coarse-grained rocks, as they expose a greater area. As the metamorphic reactions take place between the solid rock and the solutions, they require more surface exposure and hence, the rocks with fractures and joints, are more easily metamorphosed.

A change in the environment, due to an increase in the pressure and temperature, favours metamorphic reactions. This is particularly so in deep seated rocks. Burial and ocean floor metamorphism are good evidence of the importance of depth in metamorphism.

Metamorphic changes are related to progressive chemical reactions brought about by increasing temperatures and pressures. Longer the time taken by these reactions, the more potent they are in bringing about the metamorphism of the pre-existing rocks.

18.3 KINDS OF METAMORPHISM

There are two kinds of metamorphism which are recognised on the basis of the modes of occurrence of metamorphic rocks. They are (1) local metamorphism and (2) regional metamorphism.

Local metamorphism is restricted to those regions where the changes in temperature and pressure are restricted to a local area and is a short term process. **Regional** metamorphism takes place on a regional scale. Hence, it is related to large scale tectonism and magmatism, which are long term processes. These 2 kinds of metamorphism can be further divided into various types as given below.

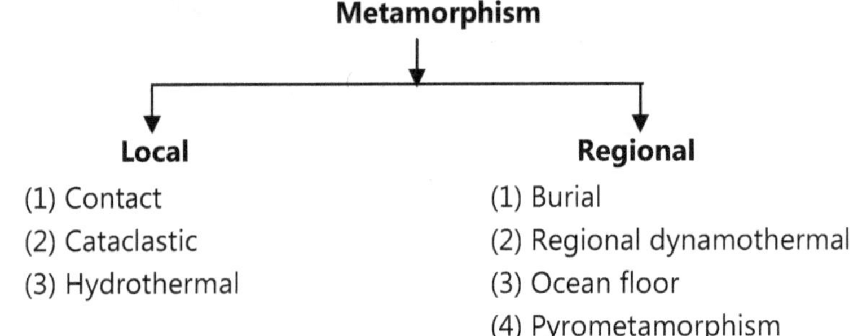

Local metamorphism: Following are different types of metamorphism which occur locally

(1) Contact metamorphism occurs when country or host rocks are intruded by magma. The high temperature of the magma heats the host rocks which then undergo contact metamorphism. This process involves sharing of the heat and the hot fluids between magma and intruded rocks. The pressure effects are almost negligible and recrystallisation is the major metamorphic effect. As pressure effects are not of much significance, the minerals do not show re-orientation.

(2) Cataclastic metamorphism takes place when directed pressure or stress with no appreciable rise in temperature leads to crushing, fracturing and granulation of the affected rock but without the formation of new mineral species. Hence, these effects are mainly structural though they may be accompanied by some mineralogical modifications (e.g., andalusite in slates). The harder minerals remain uncrushed during this metamorphic process while the less resistant minerals undergo grinding to form almost a powder. This causes the lamination of the argillaceous rocks to get deformed with the development of slaty cleavages.

(3) Hydrothermal metamorphism takes place due to the strong effect of water at high temperature, resulting in the formation of new minerals. The mineralogy of the affected rock changes with the addition or removal of water.

(4) Pyrometamorphism is a local metamorphism at very high temperature when xenoliths (foreign rock fragments) are engulfed within the magmas or more often in the lavas.

Regional metamorphism:

(1) Burial metamorphism brings about progressive mineralogical changes in the affected rock when thick sediments or thick volcanic rocks get deeply buried and are subjected to varying pressures and temperatures. The metamorphic changes are spread over a wide range, from the diagenetic changes at one end to the regional metamorphism at the other. The temperatures range from 300°C to 415°C.

(2) Regional dynamothermal: The term regional metamorphism is applied to a metamorphism which affects wide areas, predominantly the orogenic belts and continues for 100 to 150 million years. During such processes, the earth's crust experiences considerable rise in temperature and pressure. The pre-existing rocks are subjected to more or less complete recrystallisation and result in development of new structures. Foliation is characteristic feature of these rocks and the typical products are rocks like schists, gneisses, amphibolites and granulites. Due to the application of stress on the pre-existing rocks, the minerals of the rocks get deformed / reformed into an alignment of minerals perpendicular to the maximum stress direction. In most cases, the early formed foliation may be folded, disrupted and intersected by the later foliation. Hence, repeated cycles of crystallisation and deformation are possible over a long time span. Another characteristic feature of these processes is the wide variety of mineral assemblage developed from the pre-existing rocks with similar chemical composition. Hence, basalt may give rise to varied metamorphic products like chlorite schist, glaucophane schist, amphibolite or granulite.

The intensity of regional metamorphism is a function of temperature, pressure and shearing stresses, which vary according to the depth of the burial of the rocks. When argillaceous rocks like shales are subjected to dynamothermal metamorphism, newly developed micaceous minerals are oriented perpendicular to the maximum stress direction, yielding phyllites (under moderately high temperature and pressure conditions), or schists (at higher temperature and pressure conditions). Due to further increase in the pressure and temperature conditions, the phyllite is modified into a schist. At the higher grade of metamorphism, due to alternation of granulose and schistose structures, a gneissose texture is developed.

(3) Ocean floor metamorphism pertains to those metamorphic changes that take place at the ocean floor near the Mid-oceanic ridges. The temperature gradient being very high in these regions, the chemical changes are temperature dependent.

POINTS TO REMEMBER

- Pre-existing rocks that have been converted by pronounced temperature and pressure changes into a totally new rock variety, are known as metamorphic rocks.
- Metamorphism is defined as the 'mineralogical and structural adjustment of solid rocks, to physical and chemical conditions, which have been imposed at depths below the surface zones of weathering and cementation and which differ from the conditions under which the rocks in question originated'.
- Local metamorphism is restricted to those regions where the changes in temperature and pressure are restricted to a local area and is a short term process.
- Contact metamorphism is a type of local metamorphism which occurs when magma intrudes into the country or host rock.
- Cataclastic metamorphism is also a local metamorphism in which directed pressure (or stress) plays the major role, with no appreciable increase in the temperature of the affected rock.
- Hydrothermal metamorphism takes place due to the strong effect of water at high temperature, resulting in the formation of new minerals.
- Pyrometamorphism is a local metamorphism at very high temperature when xenoliths (foreign rock fragments) are engulfed within the magmas or more often in the lavas.
- Regional metamorphism takes place on a regional scale.
- Burial metamorphism is a kind of regional metamorphism which brings about progressive mineralogical changes in the affected rock when thick sediments or thick volcanic rocks get deeply buried and are subjected to varying pressures and temperatures.

- Regional dynamothermal (orogenic) metamorphism takes place mainly in the orogenic belts spread over a large region.
- Ocean floor metamorphism pertains to those metamorphic changes that take place at the ocean floor near the Mid-oceanic ridges.

EXERCISE

1. Define metamorphism. Give the classification of metamorphic rocks in a tabular form.

2. Define metamorphism. Explain in brief different factors controlling metamorphism.

3. What are different kinds of metamorphism?
   Explain local metamorphism.

4. Write notes on
   (a) Agents of metamorphism
   (b) Burial and ocean floor metamorphism
   (c) Cataclastic metamorphism
   (d) Thermal metamorphism
   (e) Dynamothermal metamorphism
   (f) Pyrometamorphism and orogenic metamorphism

Chapter **19**...

STRUCTURES OF METAMORPHIC ROCKS

Contents ...

19.1 INTRODUCTION

When a rock is subjected to different kinds of metamorphism, it gets modified, partially or completely, in its mineral composition and structure. As a result, the original texture / structure of the rock gets obliterated and a new metamorphic structure comes into being. The pattern of the arrangement of the newly formed metamorphic minerals is characteristic of the kind of metamorphism and the dominant agent of metamorphism. During metamorphic re-crystallisation, as the new minerals grow in the solid state, there is a progressive change in the shape of the minerals. In a temperature dominant metamorphic process, the mineral undergoes only re-crystallisation without mineral re-orientation. Hence, metamorphic structures are either foliated or unfoliated.

The terms foliation and schistosity are applied to the rocks whose minerals have been re-oriented with or without re-crystallisation and the structures are described as slaty, schistose and gneissose. The unfoliated rocks, in contrast, show no such preferred orientation of their re-crystallised minerals and develop a granulose texture.

The texture of the original rock which has been metamorphosed, is described by pre-fixing the texture with the term *blasto*. Hence, blasto-porphyritic texture describes a rock which shows porphyritic texture and which has developed a new structure during metamorphism, with relicts of the older porphyritic texture. But, when there is an entirely new metamorphic structure developed during metamorphism, which resembles the porphyritic texture of igneous rocks, then the term *blastic* is used for such rocks as a suffix, i.e., porphyroblastic texture. The large minerals that are developed during metamorphism are called as porphyroblasts.

The five types of structures commonly seen in the metamorphic rocks are described below (1) maculose (2) slaty (3) granulose (4) schistose and (5) gneissose.

19.2 MACULOSE STRUCTURE

It is a structure which results from the re-crytallisation of an argillaceous rock, under the effects of thermal metamorphism. The heat causes the fine grained rock to re-crystallize. The alumino-silicate constituents of the rock may grow to a larger size and give rise to large grains embedded in a fine grained groundmass. For example, in the rock chiastolite slate, the larger grains of chiastolite (andalusite having a carbonaceous impurity) are embedded in a fine grained recrystallized groundmass, as shown in Fig. 19.1.

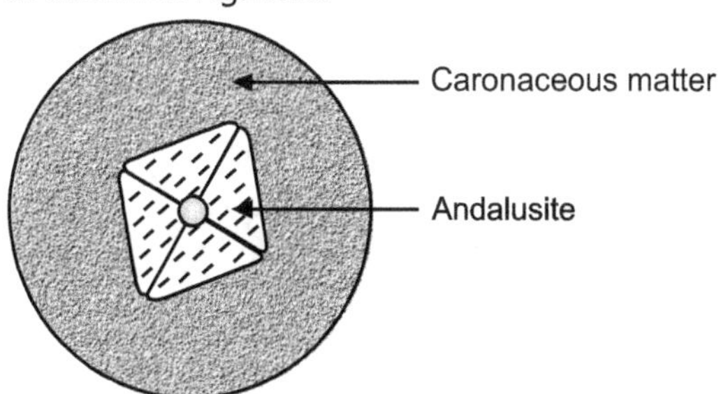

Fig. 19.1: Maculose structure in chiastolite slate

19.3 SLATY CLEAVAGE

Argillaceous sediments such as shales, under the influence of high stress (directed pressure) with low to moderate temperature, are compressed and converted to slates (Fig.19.2). Minute flakes of micaceous minerals grow with their cleavage surfaces at right angles to the direction of maximum compression. The rock, therefore, possesses a preferential direction of splitting known as slaty cleavage. Slaty cleavage often occurs after diagenesis and is the first cleavage feature to form after deformation begins. These planar structures are normally caused by the preferred orientation of mica crystals.

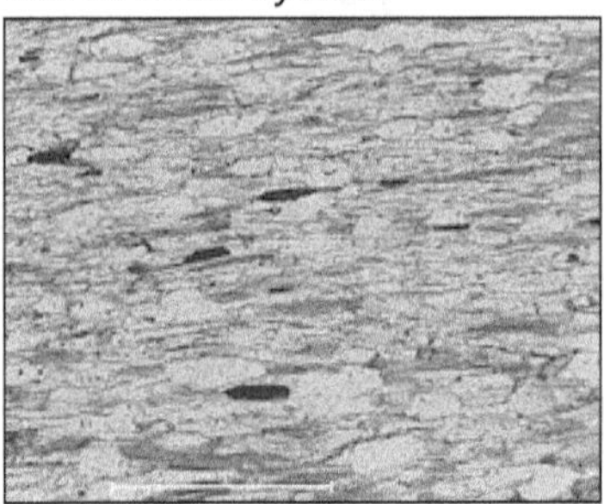

Fig. 19.2: Slaty cleavage

19.4 GRANULOSE STRUCTURE

It is a structure which is well observed in marbles and quartzites. It results from the thermal metamorphism of limestones and sandstones, respectively. The thermal metamorphism causes re-crystallisation of the grains, which develop interlocking of the grains and the development of common mineral boundaries. The finer grains of the pre-existing rocks respond to an increase in temperature by the process of re-crystallisation (Fig. 19.3).

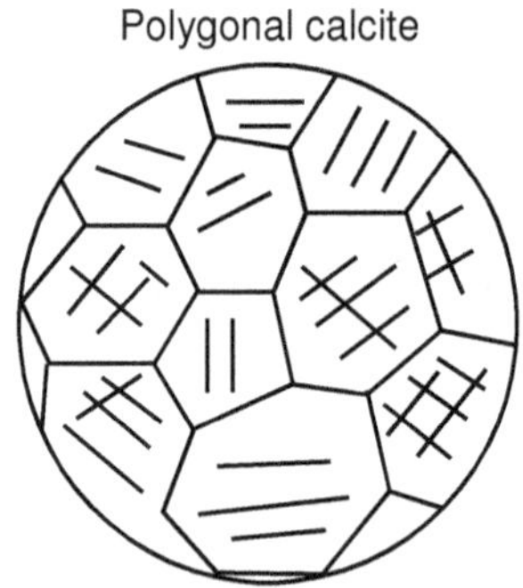

Fig. 19.3: Granulose structure in marble

19.5 SCHISTOSE STRUCTURE

When dynamo-thermal or regional metamorphism affects a pre-existing rock, it brings about the formation of new minerals and structures. The combination of stress and heat favours the development of platy or elongated minerals, which align themselves parallel to each other (Fig. 19.4).

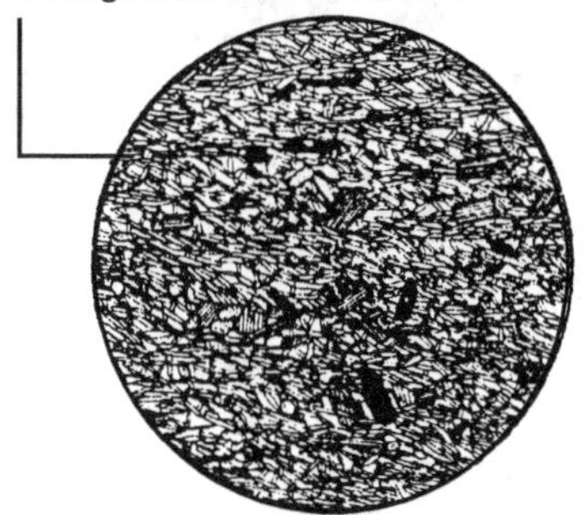

Fig. 19.4: Schistose structure

The linear arrangement of minerals in a direction perpendicular to the maximum stress direction is characteristic of these structures. Such a parallel alignment of minerals in a rock gives rise to schistose structure.

The unidirectional alignment of the minerals such as biotite, muscovite, chlorite and talc, which are flaky and highly cleavable, and hornblende, which is prismatic with a needle-like appearance, is responsible for the schistose structure. These structures are developed with high temperatures and with strong directed pressure or stress. Because of directed pressure, the minerals form layers or folia, arranged in parallel layers or bands. This parallel alignment of the flaky or platy minerals characteristically gives rise to the foliation in metamorphic rocks. A rock which has developed foliations due to parallel alignment of minerals like mica, chlorite and hornblende, shows schistose structure. The elongated and prismatic minerals like hornblende give rise to a linear schistosity, whereas flaky minerals like mica and chlorite give rise to a lamellar schistosity.

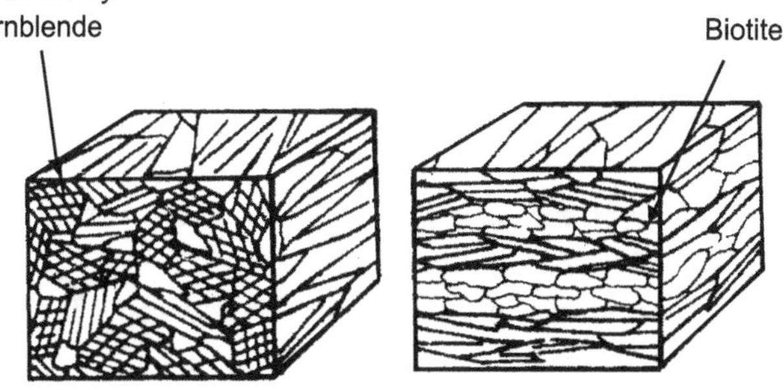

Fig. 19.5: Linear and lamellar schistosity

Basic igneous rocks subjected to this kind of metamorphism give rise to schistose rocks such as biotite schist and hornblende schist.

19.6 GNEISSOSE STRUCTURE

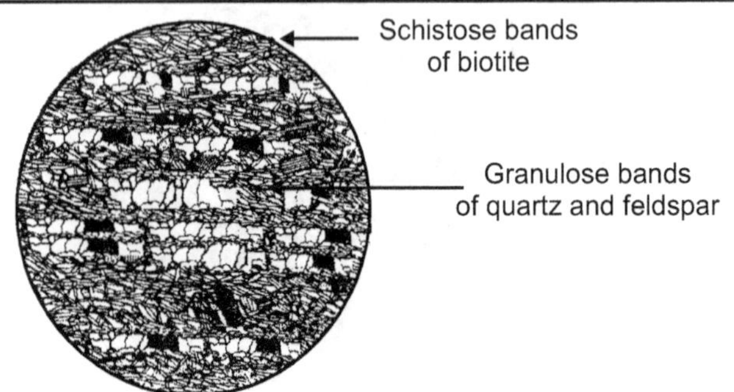

Fig. 19.6: Gneissose structure in hornblende gneiss

This structure is developed when the original rock possesses flaky and hard minerals. The flaky minerals like mica, hornblende and actinolite, give rise to a schistose structure at the end of the metamorphic process, while the hard and resistant minerals like quartz just recrystallise into a granulose structure. When both mica and quartz develop during metamorphism, the two structures (schistose and granulose), alternate with each other in a new rock, giving rise to the gneissose structure (Fig. 19.6). At a high grade of dynamo-thermal or regional metamorphism the rock acquires gneissose structure. It shows alternate layers of granulose and schistose bands. The granulose bands are made up of quartz and k-feldspars, while the schistose bands are made up of mica.

POINTS TO REMEMBER

- Blasto-porphyritic texture describes a rock which shows porphyritic texture and which has developed a new structure during metamorphism.

- The large minerals that are developed during metamorphism are called as porphyroblasts.

- Slaty cleavage often occurs after diagenesis and is the first cleavage feature to form after deformation begins.

- These planar structures are normally caused by the preferred orientation of mica crystals.

- Maculose structure results from the re-crytallisation of an argillaceous rock, under the effects of thermal metamorphism.

- Granulose structure results from the thermal metamorphism of limestones and sandstones, respectively.

- The linear arrangement of minerals in a direction perpendicular to the maximum stress direction is characteristic of schistose structures.

- Gneissose structure is developed when the original rock possesses flaky and hard minerals.

EXERCISE

1. Define metamorphism and describe the following
   (a) Granulose structure
   (b) Schistose structure
   (c) Maculose structure
   (d) Linear and lamellar schistosity
   (e) Foliation and schistosity

❖ ❖ ❖

Chapter **20**...

TABULAR CLASSIFICATION OF METAMORPHIC ROCKS

Contents ...

20.1 INTRODUCTION

The classification of metamorphic rocks is rather difficult because a particular metamorphic rock can be formed by the action of heat, pressure and chemical fluids on different pre-existing rocks. At the same time, one type of rock may give rise to a variety of metamorphic rocks under different kinds of metamorphism with varying degrees of metamorphism. For example, when argillaceous rocks (e.g., mudstone, shale or siltstone), are subjected to dynamothermal metamorphism, the low-temperature minerals like chlorite and muscovite are initially developed. But, when the temperature increases further, the high-temperature biotite is developed, at the expense of muscovite and chlorite. With still higher temperature, minerals like garnet, staurolite and kyanite are formed. The classification of metamorphic rocks, is therefore, based on the factors that cause metamorphism or the different facies and zones of metamorphism having a specific mineral assemblage. In the present text, a simplified classification is given, based on the composition of the original rock, the type of metamorphism and the metamorphic rock products (Table 20.1).

Table 20.1: Type of metamorphism and metamorphic rock products

Original Rock	Type of Metamorphism	Product of Metamorphism
Argillaceous	Cataclastic	Slate
Sandstone	Thermal	Quartzite
Limestone	Thermal	Marble
Argillaceous (shale, mudstone)	Dynamothermal	Hornblende schist, Mica schist, Hornblende gneiss
Basic igneous rock (basalt, gabbro)	Dynamothermal	Hornoblende schist, Hornblende geneiss, Biotite schist, Biotite gneiss.

20.2 DESCRIPTION OF SOME METAMORPHIC ROCKS

20.2.1 Slate

Argillaceous sediments such as shales, under the influence of high stress (directed pressure) with low to moderate temperature, are compressed and converted to slates. Minute flakes of chlorite and sericite grow with their cleavage surfaces at right angles to the direction of maximum compression, while the minerals like quartz recrystallise with their length parallel to it. The rock, therefore, possesses a preferential direction of splitting known as slaty cleavages (which is different from mineral cleavage), parallel to the flat surfaces of the oriented crystals and independent of original bedding. By virtue of their cleavage character, slates split easily into very thin sheets or slabs of uniform thickness. Cleavages in the slates develop under the influence of metamorphism and are therefore, not related to the atomic structure as in case of mineral cleavages.

20.2.2 Quartzite

Quartzite is a non-foliated metamorphic rock, composed almost entirely of quartz minerals. It is developed when a quartz-rich sandstone is altered by thermal metamorphism. High temperature conditions bring about complete recrystallization of the sand grains and the silica cement that binds them together. The result is a network of interlocking quartz grains of incredible strength. The interlocking crystalline structure of quartzite makes it a hard, tough, durable rock.

As a result, equidimensional grains of quartz, arranged on a mosaic pattern, give rise to a granulose texture. Quartzite is usually white to gray in color. Some rocks with impurities of iron give rise to pink, red, or purple quartzite. Other impurities can also impart yellow, orange, brown, green, or blue tinges to quartzite.

20.2.3 Marble

Marble is a metamorphic rock resulting from thermal metamorphism of sedimentary carbonate rocks, most commonly limestones. It is mainly composed of recrystallized calcite minerals. Marbles are developed when a limestones are subjected to very high temperature with negligible pressure acting on them. It is typically a non-foliated rock which has interlocking mosaic arrangement of polygonal calcite grains that give rise to granulose texture. Due to recrystallisation, the primary sedimentary textures and structures of the original carbonate rock are totally modified or destroyed.

When the original limestone has impurities, such as carbonaceous, iron or other minerals, then the marble appears with different colours e.g., black (carbonaceous), serpentine (green) and iron (red).

Marble is commonly used for sculpture and as a building material. Metamorphism causes variable recrystallization of the original carbonate mineral grains.

Pure white marble is the result of thermal metamorphism of a pure limestone. The characteristic veins of many colored marble varieties are usually due to various mineral impurities such as clay, silt, sand, iron oxides or chert, which were originally present as grains or layers in the limestone. Green coloration is often due to serpentine resulting from originally magnesium-rich limestone with silica impurities.

20.2.4 Hornblende Schist

Schist is a very common crystalline metamorphic rock, generally consisting of medium sized mineral grains, which can be distinguished either with naked eye or with a hand lens. These rocks are formed by the dynamo-thermal or regional metamorphism of basic igneous rocks.

A hornblende schist predominantly consists of dark hornblende minerals, alongwith few streaks of actinolite, which are arranged parallel to each other, imparting schistosity to the rock. Because of the parallel arrangement of these minerals, schists break into foliae (thin sheets) which have lustrous surfaces. The name 'schist' was originally used to denote the property of splitting into folia.

Different varieties of hornblende schists are formed due to dynamo-thermal metamorphism of sedimentary or igneous rocks under varying conditions of temperature and pressure. The hornblende schist is named after the predominant dark coloured hornblende mineral, occurring in it.

20.2.5 Mica Schist

When fine grained micaceous sedimentary rocks (argillaceous) like shales are subjected to dynamo-thermal metamorphism, mica schists are developed. Due to increased temperature during metamorphism, there occurs appreciable recrystallisation of the minerals of the pre-existing rock. The directed pressure further brings about reorientation of the newly crystallized mica flakes. Their parallel alignment gives a typical schistose structure to the metamorphic rock.

Similar to hornblende schist, the parallel arrangement of mica flakes gives rise to breaking of the rock into foliae (thin sheets) after erosion.

20.2.6 Hornblende Gneiss

This type of rock characteristically develops due to the effects of dynamo-thermal or regional metamorphism on different rocks. They show banding of light and dark colours. The dark bands are mainly composed of hornblende minerals, which are arranged parallel to each other, forming a schistose band. On the other hand, the light coloured bands are composed of quartzo-feldspathic minerals forming granulose bands. Due to alternation of schistose and granulose bands, the gneissose structure is developed. This rock is resistant to breaking as compared to a schist.

This rock is also named after the predominance of dark coloured minerals. Hence, a metamorphic rock with schistose bands of greenish black columnar needles of hornblende, and granulose bands of quartzo-feldspathic material, is known as hornblende gneiss (Fig. 20.1).

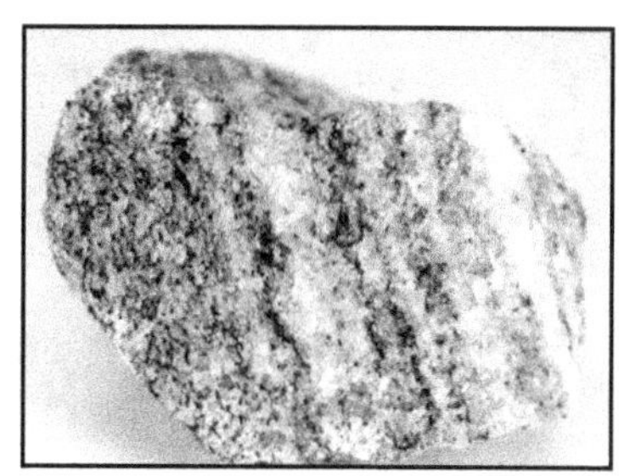

Fig. 20.1: Hornblende Gneiss

If there are large number of granulose bands (quartzo-feldspathic) and the schistose bands of hornblende needles, then it is likely that the rock is developed due to dynamothermal metamorphism of basic igneous rocks.

POINTS TO REMEMBER

- The classification of metamorphic rocks is based on the factors that cause metamorphism or the different facies and zones of metamorphism having a specific mineral assemblage.
- The contact metamorphism involves the sharing of the heat and the hot fluids between the magma and the intruded rocks.
- Cataclastic metamorphism takes place when directed pressure leads to the crushing, fracturing and granulation of the affected rock, without appreciable rise in the temperature.
- Regional metamorphism affects wide areas, predominantly the orogenic belts and continues for 100 to 150 million years.
- Slaty cleavages in the rock are parallel to the flat surfaces of the oriented crystals and independent of original bedding.
- The quartzite consists of equidimensional grains of quartz arranged on a mosaic pattern giving rise to granulose texture.
- A schist predominantly consists of dark minerals like biotite, hornblende and actinolite, which are arranged parallel to each other, imparting schistosity to the rock.
- Gneiss characteristically develops due to the effects of dynamo-thermal or regional metamorphism on different rocks. They show alternate banding of light and dark colours.

EXERCISE

1. Define metamorphism. Give the classification of metamorphic rocks in a tabular form.
2. Describe the following
    (a) Slate
    (b) Phyllite
    (c) Schist
    (d) Gneiss
    (e) Quartzite
3. What are different kinds of metamorphism? Explain any one of them in detail.

9 789389 825411